W9-AUN-783

plane trigonometry with tables

GOVERNORS STATE UNIVERSITY
UNIVERSITY PARK
IL. 60466

GOVERNORS STATE UNIVERSITY LIBRARY

3 1611 00228 0052

plane trigonometry with tables

fifth edition

gordon fuller

Professor of Mathematics, Emeritus
Texas Tech University

McGRAW-HILL BOOK COMPANY

New York   St. Louis   San Francisco   Auckland   Bogotá   Düsseldorf
Johannesburg   London   Madrid   Mexico   Montreal   New Delhi   Panama
Paris   São Paulo   Singapore   Sydney   Tokyo   Toronto

QA 533 .F85 1978

Fuller, Gordon, 1894-

Plane trigonometry, with
  tables
            266556

This book was set in Baskerville by York Graphic Services, Inc.
The editors were A. Anthony Arthur and James W. Bradley;
the designer was Anne Canevari Green;
the production supervisor was Leroy A. Young.
Von Hoffmann Press, Inc., was printer and binder

plane trigonometry
with tables

Copyright © 1978, 1972, 1966, 1959 by McGraw-Hill, Inc.
All rights reserved.
Copyright © 1950 by McGraw-Hill, Inc. All rights reserved.
Printed in the United States of America.
No part of this publication may be reproduced, stored in a retrieval system,
or transmitted, in any form or by any means, electronic, mechanical,
photocopying, recording, or otherwise, without the prior written
permission of the publisher.

1 2 3 4 5 6 7 8 9 0 VHVH 7 8 3 2 1 0 9 8 7

Library of Congress Cataloging in Publication Data

Fuller, Gordon, date
     Plane trigonometry, with tables.
     Includes index.
     1. Trigonometry, Plane.   I.   Title.
QA533.F85   1978   516′.24   77-22329
ISBN 0-07-022612-1

contents

preface

The fifth edition of "Plane Trigonometry with Tables" is well suited for students who need a sufficient mastery of trigonometry for use in analytic geometry, calculus, and more advanced mathematics. This edition retains all the topics of the fourth edition, as well as the various features which have proved to be especially appropriate. Almost all the problems are new. New also is the review exercise at the end of each chapter.

The computational problems in the book can be handled much more quickly and accurately through the use of a hand-held calculator. The efficiency and sophistication of many modern calculators is such that the traditional tables of logarithmic and trigonometric functions could well be eliminated. It is thought, however, that students should be familiar with the trigonometric and logarithmic tables; hence it is recommended that a fair number of problems be assigned which are to be solved by the use of tables.

Although largely in the nature of a review, the first chapter contains a discussion of a few important introductory concepts. The idea of sets and the notation for a set are introduced and used whenever appropriate. Directed lines and directed distances are discussed and applied in associating real numbers with the points of a line. The formula for the distance between two points is derived with attention to the special case of two points on a line parallel to a coordinate axis. Later, the distance formula has two important applications. First, it plays a key role in the proof of the addition formulas of trigonometry; and second, it is used in the derivation of the law of cosines for triangles. A function and the graph of a function are discussed and illustrated.

In the second chapter, preparatory to defining the trigonometric functions, we discuss the degree and the radian. These two units of angular measure are used throughout the book We then define the all-important trigonometric functions, showing the close relationship between the trigonometric functions of angles and the trigonometric functions of real numbers. Because of this close relationship, in the discussions which follow we have the option of considering the domains of the trigonometric functions as sets of angles or as sets of real numbers. Hence, separate treatments of these two types of functions are not necessary.

In the third chapter, approximate data and tables of natural trigonometric functions are considered. Suggestions are given for facilitating operations involving approximations, and conventions for rounding off numbers are explained. We next turn to certain right-triangle problems. Since the data in most of the problems consist of two or three significant figures, the necessary computations are not tedious, and the student can do enough problems in a short time to grasp the ideas involved and to appreciate this practical aspect of trigonometry. Here, and throughout the book, the results of worked-out computational problems and the answers to such problems are consistent with the accuracy of the given data.

Chapters 4 to 6 deal with a study of analytic trigonometry. The basic identities of trigonometry are handled in careful detail; the derivations are carried out with full explanations and are illustrated with numerous well-chosen examples. Also, suggestions are made for proving other identities.

Careful discussion is given to the line-segment representation of the values of the trigonometric functions. This representation, stemming directly from the definitions of the functions, furnishes an easy way of visualizing the variation of the values of the functions as the angle increases from 0 to 2π. The graphs of the trigonometric functions are treated with ample explanations. In addition to the graphs of the trigonometric functions of x, the graphs of the functions of bx and $bx + c$, where b and c are constants, are discussed. Also included are sums of functions of these types. Sets of real numbers are chosen as the domains in all the graphs.

Trigonometric equations and inverse trigonometric functions are then considered, and clear explanations and detailed examples are presented. The inverse functions, which are confusing to many students, are treated more thoroughly than is generally the case.

In Chapter 7, a study of logarithms precedes the study of the oblique triangle. Although many students study logarithms in algebra, a surprising number come to trigonometry with little understanding of the theory or use of logarithms. Hence, a thorough treatment of the topic, such as that found in Chapter 7, seems in order. One simple rule is developed to give the relation of the decimal point of a number to the characteristics of its logarithm.

The law of sines, the law of cosines, and the areas of triangles are treated in Chapter 8. A satisfactory coverage of the material can be made in a very few lessons by assigning problems which call for accuracy to only two or three significant figures. This will enable the student to do enough problems, in a short time, to gain an appreciation of the methods and principles involved. In addition, the experience in deriving the various formulas, or understanding the derivations, is of much value to the student.

Chapter 9 is devoted to an elementary study of vectors, including some of their properties, and also applications involving triangles. This encounter with vectors will provide a valuable study for the student.

The concluding chapter deals with complex numbers. Since many students of trigonometry have had no more than a brief introduction to complex numbers, this topic has been discussed somewhat in detail. A brief treatment of polar coordinates is also included in the chapter.

As with the text proper, meticulous care has been used in planning the exercises, each of which has an abundance of problems. Answers to two-thirds of the problems are included in the text, and the answers to one-third of the problems are given in an instructor's manual. All the necessary numerical tables are bound with the book.

If hand-held calculators have been used along the way, there will be ample time for the complete coverage of all the material of the book in a class which meets three times per week.

Special thanks go to Professor Robert Parker and to Professor Horace Woodward, both of Texas Tech University, for their valuable help with the manuscript.

Gordon Fuller

1

introductory concepts

Trigonometry had its beginnings many centuries ago. As early as the second century B.C., the Greek astronomer Hipparchus made a creditable advance toward the founding of this science by collecting and extending some of the basic ideas. Since then, and particularly in modern times, the knowledge and applications of trigonometry have increased tremendously. The earlier use of trigonometry consisted largely of computing unknown sides and angles of triangles in dealing with problems in land measurement, astronomy, and navigation. Although applications of this kind have continued through the centuries, this phase of trigonometry has been surpassed in importance by the development of other aspects of the subject. The methods and concepts of trigonometry, apart from their triangle-solving uses, have contributed to the advancement of numerous branches of mathematics. In particular, trigonometric analysis has been of great importance in the development of the physical and engineering sciences. As the student progresses, he will discover that trigonometry is vital in the structure and applications of mathematics.

1-1 Sets

The idea of a set is a fundamental concept in mathematics. Through repeated use of the word, we have come to think of a set as a collection of objects or other entities. Thus, as examples, we understand the meaning of a set of china, a set of chairs, a set of books, a set of smugglers, and so on. The meaning of the word "set" is definitely and firmly fixed in our minds; we do not attempt to define the idea.

Each object of a set is called a *member* or *element* of the set. A set may be specified by listing its elements, if possible, or by describing it in such a way that it can be determined if a given object is, or is not, a member of the set.

Although many kinds of sets occur in mathematics, we shall be particularly interested in sets of numbers and sets of angles. We mention the following examples of such sets, using a capital letter to stand for each set.

A = the set of positive integers less than 6
B = the set of integers between -5 and 0
C = the set of all numbers between 1 and 4
D = the set of all angles having measures between $0°$ and $45°$
E = the set of all angles

The numbers 1, 2, 3, 4, 5 constitute the elements of set A, and -4, -3, -2, -1 are the elements of set B. It is impossible to list the elements of the sets in the other examples because each set has infinitely many elements.

It is customary to specify the elements of a set by enclosing a description of the set within braces. If a set has only a few elements, it may be simpler just to list the elements. We specify the sets above with the following notation.

$A = \{1,2,3,4,5\}$
$B = \{-4,-3,-2,-1\}$
$C = \{x \mid 1 < x < 4\}$
$D = \{\theta \mid 0° < \theta < 45°\}$
$E = \{\theta \mid \theta \text{ is any angle}\}$

The letter x in set C stands for an arbitrary member of the set, and the Greek letter θ has a like meaning in sets D and E. We recall that the symbols $<$ and $>$ mean, respectively, "is less than" and "is greater than." The vertical bars in C, D, and E may be read "such that." Thus, in words, "C is the set of numbers x such that x is greater than 1 and less than 4." The letters x and θ in the examples are variables, as in the following definition.

Definition 1-1   *A symbol, usually a letter, which may stand for any member of a specified set of objects is called a* variable. *If the set has only one member, the symbol is called a* constant.

1-2   Directed lines and segments

A line on which one direction is defined as positive and the opposite direction as negative is called a *directed line*. Similarly, any segment of the line determined by two points and the part between is called a *directed line segment*. The chosen positive direction of the line in Fig. 1-1 is indicated by the arrowhead. The points A and B determine a directed line segment. We specify that the distance from A to B, measured in the positive direction,

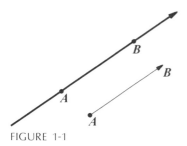

FIGURE 1-1

is positive; and the distance from B to A, measured in the negative direction, is negative. We denote these directed distances, respectively, by **AB** and **BA.** Thus, if the length of the line segment is 3, then **AB** $= 3$ and **BA** $= -3$, since distances on a directed line segment satisfy the equation

$$\mathbf{AB} = -\mathbf{BA}$$

Theorem 1-1 *If A, B, and C are three points of a directed line, then the directed distances determined by the points satisfy the equations*

$$\mathbf{AB} + \mathbf{BC} = \mathbf{AC}$$
$$\mathbf{AC} + \mathbf{CB} = \mathbf{AB}$$
$$\mathbf{BA} + \mathbf{AC} = \mathbf{BC}$$

Proof   If B is between A and C, the distances **AB**, **BC**, and **AC** all have the same sign, and **AC** is obviously equal to the sum of the other two (Fig. 1-2). The second and third equations follow readily from the first. To establish the second equation, we add $-\mathbf{BC}$ to both sides of the first equation and then use the condition that $-\mathbf{BC} = \mathbf{CB}$. Thus

$$\mathbf{AB} = \mathbf{AC} - \mathbf{BC}$$
$$\quad = \mathbf{AC} + \mathbf{CB}$$

1-3   The real number line

A basic concept of mathematics is the representation of all real numbers by the points on a directed line. The real numbers, we recall, consist of

1. The positive numbers
2. The negative numbers
3. Zero

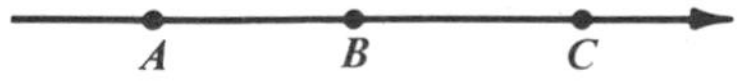

FIGURE 1-2

To establish the desired representation, we first choose a direction on a line as positive (to the right in Fig. 1-3) and select a point of the line, which we call the *origin*, to represent the number zero. Next we mark points at distances 1, 2, 3, and so on, units to the right of the origin. We let the points thus located represent the numbers 1, 2, 3, and so on. In the same way we locate points to the left of the origin to represent the numbers $-1, -2, -3$, and so on. We now have points assigned to the positive integers, the negative integers, and the integer zero. Numbers whose values are between two consecutive integers have their corresponding points between the points associated with those integers. Thus the number $2\frac{1}{4}$ corresponds to the point $2\frac{1}{4}$ units to the right of the origin. And, in general, any positive number p is represented by the point p units to the right of the origin, and a negative number $-n$ is represented by the point n units to the left of the origin. Further, we assume that every real number corresponds to one point on the line and, conversely, every point on the line corresponds to one real number. This relation of the set of real numbers and the set of points on a directed line is called a *one-to-one correspondence*.

The directed line of Fig. 1-3, with its points corresponding to real numbers, is called a *real number line* or a *real number scale*. The number corresponding to a point on the line is called the *coordinate* of the point. Since the positive numbers correspond to points in the chosen positive direction from the origin and the negative numbers correspond to points in the opposite or negative direction from the origin, we shall consider the coordinates of points on a number line to be *directed distances* from the origin. For convenience, we shall sometimes speak of a point as being a number, and vice versa. For example, we may say "the point 5" when we mean "the number 5," and "the number 5" when we mean "the point 5."

The correspondence between the set of real numbers and the points on a line furnishes a geometrical representation of the *order* property of real numbers, a topic considered in algebra. Consider a pair of numbers a and b and their corresponding points on a number line. Either the two points coincide or they do not coincide. If the points coincide, we say the numbers a and b are equal. If the point a is in the negative direction from b (to the left in Fig. 1-4), we say a is less than b. If the point a is in the positive direction from the point b, we say a is greater than b. We express the three possibilities by writing

$$a < b \qquad a = b \qquad a > b$$

when the symbol $<$ means "is less than" and the symbol $>$ means "is greater than."

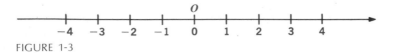

FIGURE 1-3

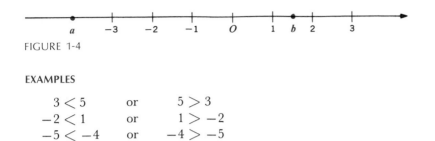

FIGURE 1-4

EXAMPLES

$$3 < 5 \qquad \text{or} \qquad 5 > 3$$
$$-2 < 1 \qquad \text{or} \qquad 1 > -2$$
$$-5 < -4 \qquad \text{or} \qquad -4 > -5$$

Frequently the concept of the absolute value of a number is of particular significance. Relative to this concept, we have the following definition.

Definition 1-2   *The* absolute value *of a real number a, denoted by* $|a|$, *is the real number such that*

$$|a| = a \text{ when a is positive or zero}$$
$$|a| = -a \text{ when a is negative}$$

According to this definition, the absolute value of every nonzero number is positive and the absolute value of zero is zero. Thus,

$$|5| = 5 \qquad |-5| = -(-5) = 5 \qquad |0| = 0$$

We obtain a geometric interpretation of the absolute value of a number by glancing at the number scale (Fig. 1-3). If a is a real number, its absolute value is the positive measurement of the distance between the origin O and the point corresponding to a. If for example $|a| = 2$, then the point corresponding to a may be two units to the right of the origin or two units to the left. And numbers whose absolute values are less than 2 are between -2 and 2; or, symbolically, if

$$|a| < 2 \qquad \text{then} \qquad -2 < a < 2$$

1-4   Rectangular coordinates

Having obtained a one-to-one correspondence between the points on a line and the system of real numbers, we next develop a scheme for putting the points of a plane into a one-to-one correspondence with a set of ordered pairs of real numbers.* This association of points and number pairs, called a *rectangular coordinate system,* was introduced in 1637 by René Descartes, a French mathematician and philosopher.

---

*Definition 1-3   *A pair of numbers (x,y) in which x is the first number and y the second number is called an* ordered pair.

We draw a horizontal line and a vertical line meeting at the origin O (Fig. 1-5). The horizontal line OX is called the *x axis* and the vertical line OY, the *y axis*. The *x* axis and the *y* axis, taken together, are called the *coordinate axes*, and the plane determined by the coordinate axes is called the *coordinate plane*. With a convenient unit of length, we make a real number line on each coordinate axis, letting the origin be the zero point. The positive direction is chosen to the right on the *x* axis and upward on the *y* axis, as indicated by the arrowheads in the figure.

If P is a point on the coordinate plane, we define the distances of the point from the coordinate axes to be *directed distances*. That is, the distance from the *y* axis is positive if P is to the right of the *y* axis and negative if P is to the left, and the distance from the *x* axis is positive if P is above the *x* axis and negative if P is below the *x* axis. Each point P of the plane has associated with it a pair of numbers called *coordinates*. The coordinates are defined in terms of the perpendicular distances from the axes to the point.

Definition 1-4   *The x coordinate, or* abscissa, *of a point P is the directed distance from the y axis to the point. The y coordinate, or* ordinate, *of a point P is the directed distance from the x axis to the point.*

A point whose abscissa is x and whose ordinate is y is designated by (x,y), in that order, the abscissa always coming first. Hence the coordinates of a point are an ordered pair of numbers. Although a pair of coordinates determine a point, the coordinates themselves are often referred to as a point.

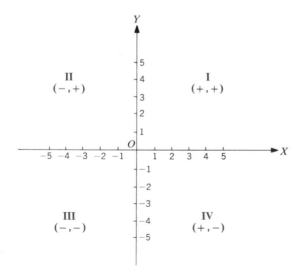

FIGURE 1-5

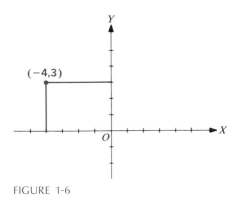

FIGURE 1-6

We assume that to any pair of real numbers (coordinates) there corresponds one definite point. Conversely, we assume that to each point of the plane there corresponds one definite pair of coordinates. This relation of points on a plane and pairs of real numbers is called a one-to-one correspondence.

A point of given coordinates is *plotted* by measuring the proper distances from the axes and marking the point thus located. For example, if the coordinates of a point are $(-4,3)$, the abscissa -4 means the point is 4 units to the left of the y axis and the ordinate 3 (plus sign understood) means the point is 3 units above the x axis. Consequently, we locate the point by going from the origin 4 units to the left along the x axis and then 3 units upward parallel to the y axis (Fig. 1-6).

The coordinate axes divide their plane into four parts, called *quadrants,* which are numbered I to IV in Fig. 1-5. The coordinates of a point in the first quadrant are both positive, which is indicated in the figure by $(+,+)$. The signs of the coordinates in each of the other quadrants are similarly indicated.

1-5   Distance between two points

The distance between two points of the coordinate plane, or the length of the line segment connecting them, can be determined from the coordinates of the points. We first observe the simple cases for the points on a horizontal line and for the points on a vertical line.

Let $P_1(x_1,y)$ and $P_2(x_2,y)$ be two points on a horizontal line, and let A be the point where the line cuts the y axis (Fig. 1-7). We have then

$$AP_1 + P_1P_2 = AP_2$$
$$P_1P_2 = AP_2 - AP_1$$
$$= x_2 - x_1$$

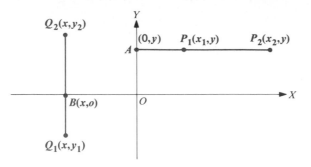

FIGURE 1-7

Similarly, for the vertical distance Q_1Q_2,

$$Q_1Q_2 = Q_1B + BQ_2$$
$$= BQ_2 - BQ_1$$
$$= y_2 - y_1$$

Hence the directed distance from a first point to a second point on a horizontal line is equal to the abscissa of the second point minus the abscissa of the first point. The distance is positive or negative according as the second point is to the right or left of the first point. A similar statement can be made about a vertical segment.

Inasmuch as the lengths of segments, without regard to direction, are often desired, we state a rule which gives results in positive quantities.

Rule *The length of a horizontal line segment joining two points is the abscissa of the point on the right minus the abscissa of the point on the left.*
The length of a vertical line segment joining two points is the ordinate of the upper point minus the ordinate of the lower point.

We apply this rule to find the lengths of the line segments in Fig. 1-8.

$$AO = 0 - (-4) = 4 \qquad BC = 4 - 1 = 3$$
$$DO = 0 - (-6) = 6 \qquad EF = 2 - (-5) = 7$$

We next consider the points $P_1(x_1,y_1)$ and $P_2(x_2,y_2)$, which determine a slanting line. Draw a line through P_1 parallel to the x axis and a line through P_2 parallel to the y axis (Fig. 1-9). These two lines intersect at the point R, whose abscissa is x_2 and whose ordinate is y_1. Hence

$$P_1R = x_2 - x_1 \qquad \text{and} \qquad RP_2 = y_2 - y_1$$

By the Pythagorean theorem (see problem 29, Exercise 1-1),

$$(P_1P_2)^2 = (x_2 - x_1)^2 + (y_2 - y_1)^2$$

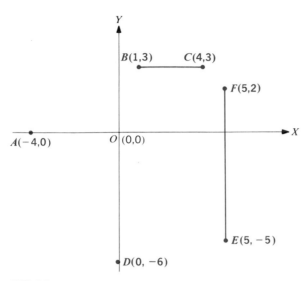

FIGURE 1-8

Denoting the length of P_1P_2 by d, we have the distance formula

$$d = \sqrt{(x_2 - x_1)^2 + (y_2 - y_1)^2}$$

We choose the positive square root because the length of a line segment is always positive. We state this formula in words.

Rule   *To find the distance between two points, add the square of the difference of the abscissas to the square of the difference of the ordinates and take the positive square root of the sum.*

In employing the distance formula, either point may be designated by (x_1, y_1) and the other by (x_2, y_2). This results from the fact that the two

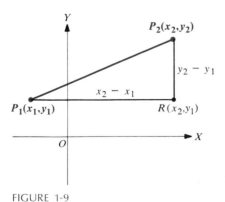

FIGURE 1-9

differences involved are squared. The square of the difference of the two numbers is unchanged when the order of subtraction is reversed.

In addition to the two coordinates of a point P, a third number, the distance of P from the origin, is associated with the point. For any point other than the origin, the distance is not zero and is defined as positive. This special distance, which we denote by r, may be expressed in terms of the abscissa x and the ordinate y of the point P. Applying the distance formula, we have

$$r = \sqrt{(x - 0)^2 + (y - 0)^2}$$

or

$$r = \sqrt{x^2 + y^2}$$

Important use of the quantities x, y, and r will soon be made.

EXAMPLE 1   The ordinate of a point is 4 and its distance r is 5. Find the abscissa of the point.

Solution   From $x^2 + y^2 = r^2$, we have

$$x^2 + 16 = 25 \qquad \text{and} \qquad x = \pm 3$$

The point may be $(3,4)$ in quadrant I or $(-3,4)$ in quadrant II.

EXAMPLE 2   Find the lengths of the sides of the triangle (Fig. 1-10) with vertices at $A(-2,-3)$, $B(6,1)$, and $C(-2,5)$.

Solution   The abscissas of A and C are the same, and therefore side AC is vertical. The length of the vertical side is the difference of the ordinates. The other sides are slanting segments, and the general distance formula

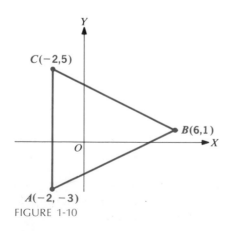

FIGURE 1-10

yields their lengths. Hence we get

$AC = 5 - (-3) = 5 + 3 = 8$

$AB = \sqrt{(6+2)^2 + (1+3)^2} = \sqrt{80} = 4\sqrt{5}$

$BC = \sqrt{(6+2)^2 + (1-5)^2} = \sqrt{80} = 4\sqrt{5}$

The lengths of the sides show that the triangle is isosceles.

Exercise 1-1

Enclose within braces a list of the elements in each of the following sets.

1. $A = \{x \mid x$ is a month of the year having 30 days$\}$

2. $B = \{x \mid x$ is one of the first three presidents of the United States$\}$

3. $C = \{x \mid x$ is an even integer between 3 and 11$\}$

Plot on coordinate paper the points in problems 4 through 9. Find the distance of each point from the origin.

4. $(4,5)$, $(13,-4)$, $(-2,-1)$ **5.** $(3,-4)$, $(4,-9)$, $(-5,-1)$

6. $(15,8)$, $(-1,-6)$, $(5,0)$ **7.** $(5,-1)$, $(15,8)$, $(-24,7)$

8. $(2,6)$, $(4,-5)$, $(6,-4)$ **9.** $(-2,-3)$, $(7,1)$, $(6,0)$

Find the missing coordinate x or y in the following problems.

10. $x = 1$, $r = \sqrt{17}$ **11.** $y = 6$, $r = \sqrt{40}$

12. $y = 4$, $r = \sqrt{241}$ **13.** $x = -8$, $y = \sqrt{233}$

14. $x = 5$, $r = 8$ **15.** $x = 9$, $r = 12$

Draw the line segment joining the two points of given coordinates in each of the following problems and find the distance between the points.

16. $(1,2)$, $(5,2)$ **17.** $(2,-3)$, $(2,3)$ **18.** $(0,1)$, $(-1,1)$

19. $(2,4)$, $(5,-6)$ **20.** $(3,7)$, $(0,3)$ **21.** $(7,8)$, $(-1,-3)$

Draw the triangle determined by the points A, B, and C in each of the following problems and find the lengths of the sides.

22. $A(5,2)$, $B(-3,-2)$, $C(-3,-4)$ **23.** $A(5,1)$, $B(3,-1)$, $C(-2,3)$

24. $A(2,3)$, $B(5,5)$, $C(9,-1)$ **25.** $A(-1,1)$, $B(8,4)$, $C(-4,7)$

26. $A(-1,0)$, $B(3,0)$, $C(2,-4)$ **27.** $A(1,1)$, $B(-1,-1)$, $C(-3,3)$

28. Plot a point $P_1(x_1,y_1)$ in the first quadrant and a point $P_2(x_2,y_2)$ in the fourth quadrant. Draw a suitable right triangle with P_1P_2 as the hypotenuse, and from the diagram *derive* the distance formula.

29. The Pythagorean theorem states that the square of the length of the hypotenuse of a right triangle is equal to the sum of the squares of the lengths of the remaining sides. Use Fig. 1-11 to prove that $c^2 = a^2 + b^2$. *Hint:* The area A of the outer square is given by $A = (a + b)^2$ and also by $A = c^2 + 4(\frac{1}{2}ab)$.

1-6   Functions and graphs

Many of the processes of mathematics are rooted in the function concept. Trigonometry, as we shall see, deals primarily with a very special type of function. First, however, we shall illustrate and define functions generally.

Suppose we consider the integers 1, 2, 3, 4, 5 and the squares of these integers 1, 4, 9, 16, 25. The first set of integers may be paired with their squares and exhibited in the form

$(1,1), (2,4), (3,9), (4,16), (5,25)$

These five ordered pairs of numbers (Sec. 1-4) furnish a simple illustration of a function.

Next let us consider the integers from 1 to 5 and all numbers between these integers. If we pair each member of this set of numbers with its square, we have infinitely many pairs of numbers. Clearly we could not write a list of all the pairs of numbers. We could, however, symbolize the pairs by writing (x,x^2) with the understanding that x may be assigned any number from 1 to 5. The resulting set of pairs of numbers (x,y), *where $y = x^2$*, constitutes another illustration of a function.

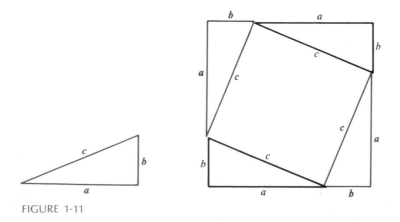

FIGURE 1-11

Definition 1-5   *If for each number x belonging to a set of numbers X there is one and only one corresponding value y, then the set of ordered pairs (x,y) is called a* function. *The set X is called the* domain *of the function and the set of y values Y is called the* range.

We emphasize that this definition requires that for any value of x of the domain there is one and only one corresponding value of y. We remark also that x and y are variables (Definition 1-1). Sometimes these variables are distinguished between by naming x the *independent variable* and y the *dependent variable*. This nomenclature seems natural because the y value in each number pair depends on the corresponding x value. Sometimes this dependence is described by saying that the dependent variable is a *function of* the independent variable.

A function may be defined by a given domain and a rule for finding the element of the range corresponding to each element of the domain. Usually the rule consists of a mathematical equation, as in the following examples.

EXAMPLE 1   Let $y = x^2 - 4x + 6$ where x may be any real number. Then, as the definition of a function requires, there is one and only one y value corresponding to each x value. Hence the equation and the specified domain give rise to a set of ordered pairs which constitute a function. The set of number pairs and the domain may be expressed by

$$\{(x,y)\,|\,y = x^2 - 4x + 6\} \qquad \text{and} \qquad \{x\,|\,x \text{ is any real number}\}$$

EXAMPLE 2   As a second illustration, suppose a body moves in a straight line for a time interval of 4 seconds according to the formula

$$s = 3t^2 + 10t$$

where the time t is in seconds and the distance s is in feet. Clearly the given conditions define an infinite set of ordered number pairs (t,s). Although we cannot list all the elements of the function, we can find the distance s which the body moves during any stated time t. Thus for integral values of t, we find, by computing, the elements $(0,0)$, $(1,13)$, $(2,32)$, $(3,57)$, $(4,88)$. The entire set of elements (number pairs) and the domain may be expressed by

$$\{(t,s)\,|\,s = 3t^2 + 10t\} \qquad \text{and} \qquad \{t\,|\,0 \le t \le 4\}$$

The second element of a number pair of a function is called the *value of the function* corresponding to the first element. Accordingly, the second elements 0, 13, 32, 57, 88 in the preceding example are values of the function corresponding to the first elements 0, 1, 2, 3, 4.

If we let a letter, f say, stand for the set of number pairs comprising a function, then the notation $f(x)$, read "f of x," denotes the value of the function at x. We warn that $f(x)$ in this context is not a product—not f times x. Although the symbol $f(x)$ represents the value of the function f corresponding to the first element x, it is sometimes convenient to let the symbol itself represent the function. Thus we shall speak loosely of "the function $f(x)$" when we really mean "the function f whose value at x is $f(x)$." For example, if $f(x) = x^2 - 4x$, or $y = x^2 - 4x$, we may regard the expression "the function $x^2 - 4x$" as a shortened form of the expression "the function f whose value at x is $x^2 - 4x$."

The following definition enables us to represent a function graphically.

Definition 1-6   *The graph of a function consists of the set of all points in a coordinate plane whose coordinates are the ordered pairs of the function. If the function is defined by an equation and a specified domain, the graph of the equation is the same as the graph of the function.*

Suppose, for example, that x may be assigned any real number in the equation

$$y = x^2 - 3x - 3$$

This equation has one and only one y value for each x value. Hence the equation with the specified values of x determines a function. Any number pair (x,y) of the function may be plotted as a point with x as the abscissa and y the ordinate. Then, according to the above definition, the set of all

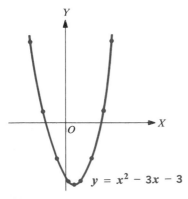

$y = x^2 - 3x - 3$

FIGURE 1-12

points thus determinable is the graph of the function. The accompanying table shows several pairs of corresponding values of x and y. The points determined by the number pairs are indicated in Fig. 1-12. These points serve as a guide for drawing a curve which approximates a part of the graph.

x	-2	-1	0	1	1.5	2	3	4	5
y	7	1	-3	-5	-5.25	-5	-3	1	7

The graph cannot be drawn completely because it extends indefinitely far into the first and second quadrants.

Chapter 1   Review exercise

1. Define the terms: *variable, directed distance, absolute value, abscissa, ordinate.*

Which, if any, of the following sets of ordered pairs of numbers satisfy the definition of a function.

2. (3,5), (4,7), (5,9)   **3.** (6,1), (5,2), (0,1), (7,2)

Find the missing coordinate x or y in the following problems.

4. $y = 5, r = 13$ **5.** $x = -4, r = \sqrt{65}$ **6.** $y = -3, r = \sqrt{58}$

P_1, P_2, and P_3 in problems 7 and 8 are vertices of a triangle. Use the distance formula and determine if each triangle is, or is not, a right triangle.

7. $P_1(4,-2), P_2(-1,3), P_3(5,9)$ **8.** $P_1(5,0), P_2(2,1), P_3(4,7)$

the trigonometric functions

2-1   Angles and units of measure

We shall devote the first part of this chapter to a discussion of angles, and consider the trigonometric functions in the remaining part. The concept of an angle is sometimes introduced as the figure formed by two half lines emanating from a common point.* The half lines are called the *sides* of the angle and the common point the *vertex*. The measure of the angle is taken as positive and equal to the amount of rotation, less than one revolution, required to bring a half line from the position of one side to the position of the other side. In trigonometry, however, the idea of an angle is enlarged to include angles of positive and negative measures of any magnitude.

In Fig. 2-1, we label one of the half lines the *initial side* and the other the *terminal side*. The curved arrow indicates a *counterclockwise* rotation from the initial side to the terminal side. For this direction of rotation, we define the angle and also the measure of the angle to be *positive*. In Fig. 2-2 the rotation from the initial side to the terminal side is *clockwise,* and we define the angle and its measure to be *negative*.

---

*A point O of a line with the part of the line extending in one direction from O is called a *half line*.

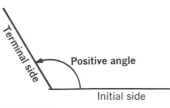

FIGURE 2-1

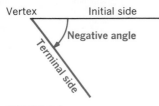

FIGURE 2-2

A half line may be rotated from coincidence with the initial side of an angle, by any chosen amount, to serve as the terminal side. Accordingly, we may visualize an angle as being formed in this way. Since the amount of rotation is arbitrary, the measure of the angle is unlimited. And the measure of the angle is positive or negative depending on whether the rotation is counterclockwise or clockwise.

There are several different units for measuring angles. The definition of an angle suggests one revolution from the position of the initial side, as indicated in Fig. 2-3, as a unit. We shall describe two other units which are used extensively.

If a rotation from the initial side to the terminal side is $\frac{1}{360}$ of a revolution, the angle is said to have a measure of one *degree* (written $1°$). One-sixtieth of a degree is called a *minute* (written $1'$), and one-sixtieth of a minute is called a *second* (written $1''$). These are the units of the *degree system* of angular measure. This system is used in astronomy, surveying, and engineering.

There is another system of angular measure, called the *radian system*, which is preferred in theoretical work, particularly in calculus. The unit of measure in this system is defined as follows:

Definition 2-1 *An angle, with its vertex at the center of a circle, which intercepts an arc equal in length to the radius of the circle is said to have a measure of 1 radian.**

Angles of 1 radian and $-2\frac{1}{2}$ radians are constructed in Fig. 2-4. The curved arrow points in the direction of rotation in each case.

The ratio of the circumference of a circle to its diameter is denoted by the Greek letter pi (π), whose value is approximately 3.1416. The definition

---

*The part of a circle between any two of its points is called an *arc* of the circle. The two points of course separate the circle into two arcs.

FIGURE 2-3

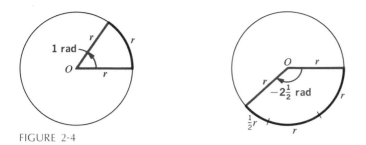

FIGURE 2-4

of π leads to the formula $c = 2\pi r$, where c stands for the circumference and r for the radius. Hence, a length of one radius can be measured off along the circumference 2π times without overlapping. This means that the entire circumference subtends a central angle of 2π, or about 6.2832, radians. It should be noticed that 2π radians are obtained regardless of the length of the radius; hence, the radian is a fixed unit of angular measure.

We may find a simple relation among the radius r of a circle, a central angle of measure θ, and the length s of the intercepted arc (Fig. 2-5). Since a central angle of measure 1 radian intercepts an arc equal in length to the radius of the circle, a central angle of measure θ radians intercepts an arc of θ times the radius; that is,

$$s = r\theta$$

This relation among the quantities s, r, and θ may be used to compute the value of any one of them if the values of the other two are known. It is essential in applying the formula that the radius and arc be measured in the same linear units, and that θ be the measure, *in radians,* of the central angle. When the radius has a measure of one unit, we note that the formula $s = r\theta$ becomes simply $s = \theta$. Later, we shall make important use of this.

We can see from our discussion of angles that an angle and its measure are two distinct entities. Notwithstanding this fact, however, mathematicians quite generally use "angle" and "measure of an angle" interchangeably. For example, if θ stands for an angle of measure $60°$, we write $\theta = 60°$, when

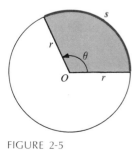

FIGURE 2-5

really the measure of the angle is equal to 60°. Since the entire circumference of a circle subtends a central angle of 2π units in terms of radians and 360 units in terms of degrees,

2π radians $= 360°$

and

π radians $= 180°$

By this relation, a radian may be expressed in terms of degrees and a degree in terms of radian measure. Thus

$$1 \text{ radian} = \frac{180°}{\pi} \approx 57.296° \approx 57° \; 17.7'†$$

and

$$1° = \frac{\pi}{180} \text{ radian} \approx 0.017453 \text{ radian}$$

When an angle is expressed in radians, the word "radian" is frequently omitted. Thus $\pi = 180°$ and $\pi/4 = 45°$ are written with the understanding that the radian is the unit of measure in the left members of these equations.

EXAMPLE 1   Express 160° in terms of π radians.

Solution   Since $1° = \pi/180$ radian, we have

$$160° = 160 \frac{\pi}{180} = \frac{8\pi}{9}$$

EXAMPLE 2   Change 91° 48' to radian measure.

Solution   We first express 48' as a decimal part of a degree and then use the approximation $1° = 0.01745$ radian. Thus

$$91° \; 48' = 91.8° = 91.8(0.01745) = 1.602 \text{ radians}$$

EXAMPLE 3   Express $5\pi/6$ in terms of degrees.

Solution   Using the relation $\pi = 180°$, we write

$$\frac{5\pi}{6} = \frac{5}{6}(180°) = 150°$$

EXAMPLE 4   Express 3.2 radians in terms of degrees and minutes.

Solution   From the relation 1 radian $\approx 57.296°$, we get

$$3.2(57.296°) \approx 183.347° \approx 183° \; 20.8'$$

---

†The symbol $\approx$ means "is approximately equal to."

EXAMPLE 5   Find the central angle subtended by an arc of length 5 if the radius is 2.

Solution   Substituting in the formula $s = r\theta$, we obtain

$5 = 2\theta$ and $\theta = 2.5$ radians $\approx 143° \; 14.4'$

EXAMPLE 6   A central angle of 38° intercepts an arc of 3 feet. Find the radius of the circle.

Solution   In order to apply the relation $s = r\theta$, we first express 38° in radian measure. Thus

$38° \approx 38(0.01745) \approx 0.663$ radian

Hence,

$3 \approx 0.663r$ and $r \approx 4.52$ feet

Exercise 2-1

Express each of the following angles in terms of π radians.

1. 20°	**2.** 90°	**3.** 270°	**4.** 300°
5. 315°	**6.** 240°	**7.** 135°	**8.** 210°
9. 330°	**10.** 80°	**11.** 225°	**12.** 140°
13. −70°	**14.** −21°	**15.** −27°	**16.** −72°

Express each angle in radian measure using the approximation $1° = 0.01745$.

17. 10°	**18.** 20°	**19.** 8° 30′	**20.** 80° 42′
21. 70° 36′	**22.** 32° 48′	**23.** −(62° 12′)	**24.** −(72° 18′)

Change each angle from radian measure to degrees.

25. $\dfrac{7\pi}{6}$	**26.** $\dfrac{4\pi}{3}$	**27.** $\dfrac{3\pi}{4}$	**28.** $\dfrac{2\pi}{9}$
29. $-\dfrac{3\pi}{6}$	**30.** $\dfrac{2\pi}{15}$	**31.** $\dfrac{3\pi}{10}$	**32.** $\dfrac{\pi}{30}$
33. $\dfrac{7\pi}{12}$	**34.** $\dfrac{5\pi}{24}$	**35.** $-\dfrac{2\pi}{15}$	**36.** $-\dfrac{7\pi}{18}$

Change each angle from radian measure to degrees, expressing a fraction of a degree in minutes. First use the approximation 1 radian = 57.296°.

37. −2	**38.** −3	**39.** 1.2	**40.** −0.06
41. 0.4	**42.** 1.51	**43.** −0.22	**44.** 5.1

45. Find the length of the arc of a circle of radius 28 inches which subtends a central angle of (*a*) 0.6 radian, (*b*) 0.25π, (*c*) 75°.

46. An arc of 16 inches subtends central angles of 1.4 radians, 0.2π, and 20°, respectively, in each of three circles. Find the radii of the circles.

47. Find the central angle in radians which intercepts an arc of 5 feet if the radius of the circle is (*a*) 10 feet, (*b*) 42 inches, (*c*) 1 yard.

48. If two cities have the same longitude and their latitudes are 40° N and 20° N, find the distance between the cities. Use 4000 miles as the radius of the earth.

49. Do problem 48 if the latitudes are 35° N and 25° S.

50. Two cities are located on the earth's equator and their longitudes differ by 170°. Figuring the radius of the earth as 4000 miles, find the distance between the cities.

51. The pendulum of a clock swings through an angle of 10°. Find the distance which the tip travels in one swing if the length of the pendulum is 30 inches.

52. Find the angle through which a pendulum swings if its length is 36 inches and its tip describes an arc of (*a*) 6 inches, (*b*) 8 inches, (*c*) 12 inches.

53. The minute hand of a clock is 3 inches long. How far does the tip of the hand travel in (*a*) 6 minutes, (*b*) 20 minutes, (*c*) 30 minutes?

54. The moon's diameter subtends an angle of about 31.1′ from the center of the earth. Taking the distance from the earth to the moon as 239,000 miles, find the diameter of the moon.

2-2   Angle in standard position

An angle which has its vertex at the origin and its initial side along the positive *x* axis is said to be in *standard position*. The location of the terminal side is determined by the magnitude and sign of the angle. The angle is usually spoken of as being *in the quadrant* which contains the terminal side. For example, we say that 135° is in quadrant II, and −140° is in quadrant III. The angles, of course, really *terminate* in those quadrants (Fig. 2-6).

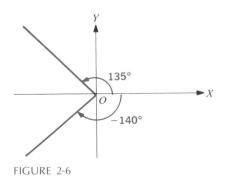

FIGURE 2-6

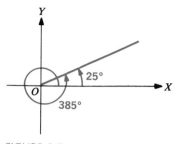

FIGURE 2-7

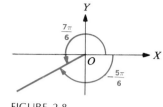

FIGURE 2-8

Angles in standard position and having the same terminal sides are said to be *coterminal*. Two coterminal angles are not necessarily equal. Thus, drawn in standard position, the pair of angles 25° and 385° are coterminal (Fig. 2-7) and the pair $7\pi/6$ and $-5\pi/6$ are coterminal (Fig. 2-8). An angle coterminal with a given angle may be obtained by adding any integral multiple of 360° or 2π. Hence, an angle has an unlimited number of coterminal angles.

EXAMPLE   Find two other angles of smallest magnitude which are coterminal with $-225°$.

Solution   Referring to Fig. 2-9, we see that 135° and 495° are the desired angles.

Exercise 2-2

Construct each of the following angles in standard position and draw a curved arrow to indicate the direction of rotation.

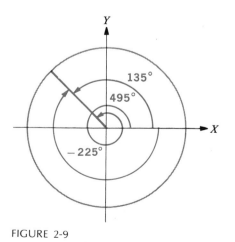

FIGURE 2-9

1. $-60°, 48°, -210°, 135°$ **2.** $-120°, 190°, -320°, 180°$
3. $95°, 325°, -65°, -330°$ **4.** $800°, -400°, 270°, -270°$

Let each of the following angles be in standard position and name the quadrant in which the terminal side lies.

5. $97°, 340°, -72°, -320°$ **6.** $-154°, -540°, 230°, 320°$

7. $\dfrac{2\pi}{3}, \dfrac{5\pi}{4}, -\dfrac{7\pi}{6}, -\dfrac{11\pi}{3}$ **8.** $-\dfrac{3\pi}{5}, -\dfrac{17\pi}{4}, \dfrac{7\pi}{6}, \dfrac{13\pi}{3}$

Find two other angles of smallest magnitude which are coterminal with each of the following angles. Assume that the angles are in standard position.

9. $70°$ **10.** $-165°$ **11.** $-230°$ **12.** $230°$
13. $-325°$ **14.** $350°$ **15.** -5π **16.** 5π

Find two other angles of smallest magnitude, one positive and one negative, which are coterminal with each of the following angles. Assume that the angles are in standard position.

17. $630°$ **18.** $-630°$ **19.** $-135°$ **20.** $-30°$
21. $580°$ **22.** $-580°$ **23.** $290°$ **24.** $420°$

2-3   Definition of the functions

In Definition 1-5 we specified that the domain and range of a function consist of numbers. In a broader sense, however, the domain or range, or both, may be objects other than numbers. The trigonometric functions, which we are now ready to define, are illustrations of functions whose domains are sets of angles.

Let θ be an angle in standard position (Fig. 2-10) and let $P(x, y)$ be any point, other than the origin, on the terminal side. Then the abscissa x, the ordinate y, and the distance r have six possible ratios called *trigonometric ratios*. Using these ratios, we define the trigonometric functions by the following equations:

$$\text{sine } \theta = \frac{\text{ordinate}}{\text{distance}} = \frac{y}{r} \qquad \text{cosecant } \theta = \frac{\text{distance}}{\text{ordinate}} = \frac{r}{y}$$

$$\text{cosine } \theta = \frac{\text{abscissa}}{\text{distance}} = \frac{x}{r} \qquad \text{secant } \theta = \frac{\text{distance}}{\text{abscissa}} = \frac{r}{x}$$

$$\text{tangent } \theta = \frac{\text{ordinate}}{\text{abscissa}} = \frac{y}{x} \qquad \text{cotangent } \theta = \frac{\text{abscissa}}{\text{ordinate}} = \frac{x}{y}$$

Each of these equations, as the ensuing discussions will reveal, determines a set of pairs of elements which conform to the idea of a function (Definition 1-5).

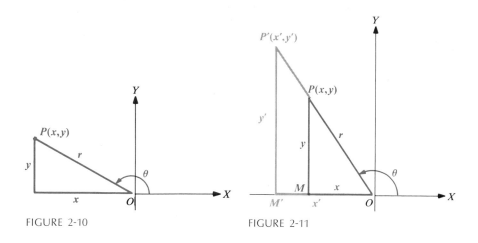

FIGURE 2-10   FIGURE 2-11

The abbreviations $\sin \theta$ for sine θ, $\cos \theta$ for cosine θ, $\tan \theta$ for tangent θ, $\cot \theta$ for cotangent θ, $\sec \theta$ for secant θ, and $\csc \theta$ for cosecant θ are customarily used.

An angle in standard position with its terminal side along a coordinate axis is called a *quadrantal angle*. One of the coordinates of points on the terminal side of a quadrantal angle is equal to zero, and consequently not all of the six trigonometric ratios exist. If the terminal side is along the y axis, the x coordinate is equal to zero. This means that the ratios y/x and r/x do not exist and, for such an angle, the tangent and secant functions are not defined. Similarly, the ratios x/y and r/y do not exist at points on the x axis. For this case the cotangent and cosecant functions are not defined.

2-4 Immediate consequences of the definitions

The values of the six ratios appearing in the definitions of the trigonometric functions do not depend on the position of the point P on the terminal side. To show that this is true, let P' be another point on the terminal side with coordinates (x', y') and distance r' (Fig. 2-11). The right triangles OMP and $OM'P'$, called *reference triangles* for θ, are similar. That is, the corresponding sides are proportional. Hence we have the equal ratios

$$\frac{y}{r} = \frac{y'}{r'}$$

which show that the same value is obtained for $\sin \theta$ by using either P or P'. Similarly, the value of each of the other functions is independent of the point selected on the terminal side.

We see now that the value of each trigonometric ratio depends solely on the position of the terminal side of the angle. Further, corresponding to an angle θ, each defined trigonometric ratio is equal to a unique real number. Because of this uniqueness, we conclude that each of the six equations of the previous section gives rise to a set of ordered pairs of elements which constitute a function. The domain of the sine and cosine functions is the set of all angles, but certain quadrantal angles are excluded from the domain of each of the other functions. As examples, the sine function comprises the set

$$\left\{ (\theta, \sin \theta) \,\middle|\, \sin \theta = \frac{y}{r} \right\}^{\dagger}$$

and the tangent function the set

$$\left\{ (\theta, \tan \theta) \,\middle|\, \tan \theta = \frac{y}{x} \text{ and } x \neq 0 \right\} \quad .$$

Note that the condition $x \neq 0$ excludes all angles with the terminal side along the y axis.

As we have observed, the value of each trigonometric ratio depends solely on the position of the terminal side of the angle. But the position of the terminal side is determined by the measure of the angle. Because of this fact, we shall equate a trigonometric function of an angle to the function of the measure of the angle. For example, if θ is an angle of measure $60°$, we write

$$\sin \theta = \sin 60°$$

Since the value of a trigonometric function of an angle in standard position is determined solely by the position of the terminal side, we see that the function has the same value for any two coterminal angles. This is illustrated by the equation

$$\sin 15° = \sin (15° + 360°)$$

and, more generally, by

$$\sin \theta = \sin (\theta + n \cdot 360°)$$

where n is any integer.

Recalling that the reciprocal of any nonzero number a is $1/a$, we discover that the six trigonometric ratios form three pairs of reciprocals, namely, y/r and r/y, x/r and r/x, y/x and x/y. Since the product of each pair of reciprocals is unity, we have the equations

$$\sin \theta \csc \theta = 1 \qquad \cos \theta \sec \theta = 1 \qquad \tan \theta \cot \theta = 1$$

---

†As we see here, $\sin \theta$ is the value of the sine function corresponding to the element θ of the domain. Although $\sin \theta$ is a function value, many writers refer to $\sin \theta$ itself as a function. We may justify this usage by considering "the function $\sin \theta$" as an abbreviation for "the sine function whose value at θ is $\sin \theta$."

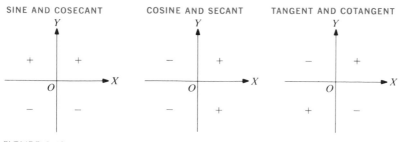

SINE AND COSECANT   COSINE AND SECANT   TANGENT AND COTANGENT

FIGURE 2-12

These equations may be reduced to the equivalent forms

$$\sin \theta = \frac{1}{\csc \theta} \qquad \cos \theta = \frac{1}{\sec \theta} \qquad \tan \theta = \frac{1}{\cot \theta}$$

$$\csc \theta = \frac{1}{\sin \theta} \qquad \sec \theta = \frac{1}{\cos \theta} \qquad \cot \theta = \frac{1}{\tan \theta}$$

Each of these equations is true for all angles except certain quadrantal angles previously mentioned. For this reason, as we shall learn in Chap. 4, the equations are called *identities*.

The distance r appearing in the definitions of the trigonometric functions is always positive, and consequently the signs of the values of the functions depend on the signs of x and y. Accordingly, $\sin \theta$ and $\csc \theta$, taking the sign of y, are positive if θ is in quadrant I or quadrant II and negative if θ is in quadrant III or quadrant IV. The pair $\cos \theta$ and $\sec \theta$ have the same sign as x and are positive if θ is in quadrant I or quadrant IV and negative if θ is in quadrant II or quadrant III. The ratios y/x and x/y are positive if x and y have like signs and negative if x and y have unlike signs. Hence, $\tan \theta$ and $\cot \theta$ are positive if θ is in quadrant I or quadrant III and negative if θ is in quadrant II or quadrant IV.

The signs of the values of the trigonometric functions for the various quadrants are indicated in Fig. 2-12. However, one should be able to visualize the terminal side of an angle in any of the four quadrants and, from the signs of x and y, give immediately the signs of the values of the functions.

EXAMPLE 1   Write the values of the functions of an angle whose terminal side passes through $(-4, -3)$.

Solution   We substitute $x = -4$ and $y = -3$ in the relation $r^2 = x^2 + y^2$ and find $r = 5$. Then we apply the definitions of the functions to the angle (Fig. 2-12) and obtain

$$\sin \theta = \frac{y}{r} = \frac{-3}{5} = -\frac{3}{5} \qquad \csc \theta = \frac{r}{y} = \frac{5}{-3} = -\frac{5}{3}$$

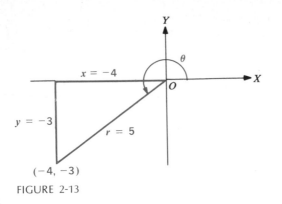

FIGURE 2-13

$$\cos \theta = \frac{x}{r} = \frac{-4}{5} = -\frac{4}{5} \qquad \sec \theta = \frac{r}{x} = \frac{5}{-4} = -\frac{5}{4}$$

$$\tan \theta = \frac{y}{x} = \frac{-3}{-4} = \frac{3}{4} \qquad \cot \theta = \frac{x}{y} = \frac{-4}{-3} = \frac{4}{3}$$

EXAMPLE 2   Find the values of the functions of 270°.

Solution   The terminal side of 270° is along the negative y axis, and for convenience we select the point $(0,-1)$ on this side (Fig. 2-14). The distance r for this point is 1, and applying the definitions of the trigonometric functions, we obtain

$$\sin 270° = \frac{-1}{1} = -1 \qquad \cot 270° = \frac{0}{-1} = 0$$

$$\cos 270° = \frac{0}{1} = 0 \qquad \csc 270° = \frac{1}{-1} = -1$$

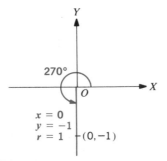

FIGURE 2-14

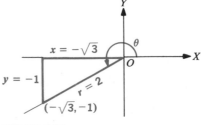

FIGURE 2-15

Notice that we have omitted $\tan 270°$ and $\sec 270°$. The ratios y/x and r/x, corresponding to the tangent and secant functions, have zero for the denominator. But division by zero is forbidden, and therefore $270°$ has no tangent and no secant. We remark that the values of the four defined functions of $270°$ are the same for all angles coterminal with $270°$. Thus, for example,

$$\sin 270° = \sin (270° + n \cdot 360°)$$

where n is any integer.

We now show in tabular form the function values of $0°$, $90°$, $180°$, and $270°$. The values for each angle should be verified by the use of a figure.

	sin	cos	tan	cot	sec	csc
0°	0	1	0	none	1	none
90°	1	0	none	0	none	1
180°	0	−1	0	none	−1	none
270°	−1	0	none	0	none	−1

As this table shows, the values of two of the functions of a quadrantal angle are 0, the values of two are either 1 or -1, and two of the functions do not exist.

EXAMPLE 3   Given $\csc \theta = -2$ and θ in quadrant III, find the values of the other functions.

Solution   Here, $r/y = -2$, and y must be negative. If we select the point on the terminal side for which $r = 2$, then y must have the value -1. To find x, we substitute these values for r and y in the relation $x^2 + y^2 = r^2$ and find

$$x^2 + (-1)^2 = 2^2 \qquad x^2 = 3 \qquad x = \pm\sqrt{3}$$

In order to obtain a third-quadrant angle, as specified, we choose $x = -\sqrt{3}$ and draw the terminal side through $(-\sqrt{3}, -1)$. Then from Fig. 2-15, we write.

$$\sin \theta = -\tfrac{1}{2} \qquad\qquad \csc \theta = -2$$

$$\cos \theta = -\frac{\sqrt{3}}{2} \qquad\quad \sec \theta = -\frac{2}{\sqrt{3}}$$

$$\tan \theta = \frac{1}{\sqrt{3}} \qquad\qquad \cot \theta = \sqrt{3}$$

EXAMPLE 4   If $\sin \theta = 0$, find the values of the other functions of θ.

Solution   We have $y/r = 0$, and hence $y = 0$. This means that the terminal side must lie on the positive x axis or the negative x axis. The angle
θ, therefore, is equal to $0°$ or $180°$ (or any angle coterminal with one of
these). The values of the other defined functions are

$$\cos 0° = 1 \qquad \cos 180° = -1$$

$$\tan 0° = 0 \qquad \tan 180° = 0$$

$$\sec 0° = 1 \qquad \sec 180° = -1$$

Exercise 2-3

The terminal side of an angle θ in standard position passes through
the given point in each of the following problems. From a point on
the terminal side drop a perpendicular to the x axis to form a
reference triangle for θ. Draw a curved angle from the initial side to
the terminal side, and give the values of the six trigonometric
functions of θ.

1. $(-\sqrt{3}, 1)$ 2. $(-\sqrt{6}, -\sqrt{3})$ 3. $(15, -8)$
4. $(-4, 3)$ 5. $(5, -12)$ 6. $(12, -5)$
7. $(4, -5)$ 8. $(\sqrt{7}, \sqrt{2})$ 9. $(\sqrt{11}, \sqrt{5})$
10. $(-5, -7)$ 11. $(-3, -5)$ 12. $(-8, 3)$

Name the quadrant or quadrants in which the terminal side of an
angle in standard position may be and satisfy the given condition or
conditions.

13. The sine is positive.   14. The cosine is negative.
15. The tangent is negative.   16. The cotangent is positive.
17. $\sin \theta < 0$, $\cos \theta > 0$ 18. $\cos < 0$, $\tan > 0$
19. $\cos \theta > 0$, $\cot \theta < 0$ 20. $\sin \theta > 0$, $\cos \theta > 0$

Give the signs of the values of the trigonometric functions of each of
the following angles.

21. $-118°$ 22. $240°$ 23. $-95°$ 24. $350°$
25. $520°$ 26. $-400°$ 27. $-600°$ 28. $900°$

29. $-\dfrac{4\pi}{3}$ 30. $\dfrac{5\pi}{3}$ 31. $\dfrac{5\pi}{11}$ 32. $\dfrac{11\pi}{3}$

Construct an angle θ in the interval $0° \le \theta < 360°$ which satisfies the
condition in each problem 33 through 42. Then write the values of
the other defined trigonometric functions of θ.

33. $\cos \theta = \frac{3}{5}$, θ in Q IV

34. $\sin \theta = \frac{5}{13}$, θ in Q II

35. $\cot \theta = \frac{8}{15}$, θ in Q III

36. $\tan \theta = -\frac{24}{7}$, θ in Q II

37. $\sec \theta = \sqrt{17}$, $\csc \theta < 0$

38. $\csc \theta = \sqrt{10}$, $\sec \theta > 0$

39. $\sin \theta = -1$

40. $\cos \theta = 1$

41. $\csc \theta = 1$

42. $\sec \theta = -1$

Write the values of the defined functions of each angle.

43. 6π **44.** 5π **45.** $-\dfrac{7\pi}{2}$ **46.** $\dfrac{3\pi}{2}$

Construct two angles in standard position which have the given function value and different terminal sides, and write the values of the other trigonometric functions of each angle.

47. $\cos \theta = \frac{9}{41}$

48. $\sin \theta = \frac{24}{25}$

49. $\tan \theta = -\frac{7}{24}$

50. $\sec \theta = \frac{29}{21}$

51. $\cot \theta = -4$

52. $\csc \theta = 2$

53. $\sin \theta = -\frac{2}{3}$

54. $\cos \theta = -\frac{1}{2}$

55. $\tan \theta = \frac{5}{3}$

2-5 Trigonometric functions of numbers

We have defined the trigonometric functions so that the domain of each function is a set of angles and the range a set of real numbers. We shall now discuss a modification of the definitions so that the domains as well as the ranges are sets of real numbers.

The length s of the arc of a circle intercepted by a central angle of θ radians is equal to the product of θ and the radius (Sec. 2-1); that is, $s = r\theta$. If the radius is chosen as the unit of length, this equation reduces to $s = \theta$. Hence, the number of radians in the central angle is equal to the number of units of length of the intercepted arc. This equality suggests that we might seek an interpretation of the trigonometric functions in which the domains are real numbers instead of angles (or measures of angles). Inasmuch as angles have positive, zero, and negative measures, it seems desirable that the domains of the new type of functions should be similarly inclusive. To achieve this end, we refer to Fig. 2-16, in which $r = 1$ and $\theta = s$. Now suppose we let s be a distance directed along the circle measured from $A(1,0)$ with the positive direction counterclockwise and the negative direction clockwise. Then, for example, we may interpret $\sin s$ as the sine of a directed distance and take the value of $\sin s$ to be the same as the value of $\sin (s$ radians). With this understanding, the sine function consists of a set of number pairs of the form $(s, \sin s)$ in which the domain is the set of all real numbers. So we may write

$$\sin s = \sin (s \text{ radians}) = \sin \left(\frac{180s}{\pi}\right)^{\circ}$$

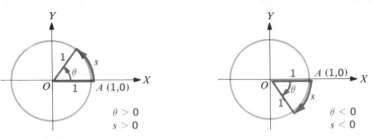

FIGURE 2-16

The preceding discussion motivates us to make the following definition:

Definition 2-2   *If θ is an angle of s radians, then any trigonometric function of θ is equal to the same-named function of s.*

On the basis of this definition and the fact that we equate a trigonometric function of an angle to the same function of the measure of the angle, we have

$$\sin \theta = \sin (s \text{ radians}) = \sin s$$
$$\cos \theta = \cos (s \text{ radians}) = \cos s$$
$$\tan \theta = \tan (s \text{ radians}) = \tan s$$

and so on, for the remaining functions.

These new functions, being introduced by use of the unit circle, are sometimes called *circular trigonometric functions* or just *circular functions*.

Clearly, the trigonometric functions and the circular functions differ only in their domains. For example, the ordered pair $(\theta, \sin \theta)$ corresponds to the ordered pair $(s, \sin s)$, where s is the number of radians in the angle θ and $\sin \theta = \sin s$. As a consequence of this close tie between the two types of functions, a property of one of the types has its counterpart in the other type. We shall be studying formulas and equations involving trigonometric function values in which the domain may be regarded either as a set of angles or a set of real numbers.

As the student progresses, he will discover that trigonometric functions of numbers are used freely and effectively in advanced mathematics, especially in calculus. This does not mean, however, that the trigonometric functions of angles are to be supplanted by the trigonometric functions of numbers. In some instances, it is advantageous to use numbers as the domain of a function, and in others, the problem is better handled with angles as the domain.

Chapter 2   Review exercise

Define the following terms.

1. radian   **2.** degree   **3.** standard position of an angle
4. coterminal angles   **5.** quadrantal angles

Express each angle in terms of π radians.

6. $135°$ **7.** $330°$ **8.** $90°$ **9.** $-40°$

Express each angle in radian measure, using the approximation $1° = 0.017$.

10. $50°$ **11.** $10° \, 30'$ **12.** $-(6° \, 50')$ **13.** $-(65° \, 12')$

Change each angle from radian measure to degrees, expressing a fraction of a degree in minutes.

14. 4 **15.** -2.3 **16.** 0.8 **17.** 1.54

18. The pendulum of a clock swings through an angle of $12°$. Find the distance which the tip travels in one swing if the length of the pendulum is 40 inches.
19. The six trigonometric functions of an angle are defined in terms of the quantities x, y, and r. What do x, y, and r represent in the definition?

The terminal side of an angle in standard position passes through the given point in each of the following problems. Find the values of the six trigonometric functions of the angle.

20. $(3,4)$ **21.** $(-5,12)$ **22.** $(-\sqrt{5}, \sqrt{11})$ **23.** $(-\sqrt{3}, -\sqrt{6})$

24. If $\cos \theta = 1/\sqrt{10}$, θ in Q IV, write the values of the other trigonometric functions of θ.
25. If $\sec \theta = \sqrt{10}$, $\csc \theta > 0$, write the values of the other functions of θ.

Construct two angles in standard position which have the given function value and different terminal sides. Then write the values of the other trigonometric functions of each angle.

26. $\sin \theta = \frac{9}{41}$ **27.** $\cos \theta = -\frac{3}{4}$ **28.** $\tan \theta = \frac{5}{6}$

29. Which trigonometric function values are never greater than 1 nor less than -1?

tables of trigonometric functions and solutions of right triangles

3-1   Operations with approximate data

The results of physical measurements are only approximately correct. The degree of accuracy in a measurement depends, among other things, on the quality of the measuring instrument and the aptness of the observer. The weight of an adult is usually expressed as an integer. Thus, we might give a man's weight as 174 pounds if he weighs between 173.5 and 174.5 pounds. We then consider 174 as an approximation to the nearest integer.

The exact area of the surface of the earth is not known. It has been computed to the nearest million to be 197,000,000 square miles. This means the exact area is between 196,500,000 and 197,500,000 square miles. The first three digits in 197,000,000 are correct (the 7 may be in error not to exceed $\frac{1}{2}$ in its place); the zeros are used to fill in the places of unknown digits up to the position of the decimal point. It is common practice to use only zeros for this purpose. Zeros are also used to locate the decimal point to the left of a set of digits. Suppose, for example, the length of an object is measured to be 0.0041 meter. The zeros in this number serve only to fix the position of the decimal point; they do not indicate accuracy of measurement. In each of the two preceding measurements the zero digits and the nonzero digits have different purposes, and we distinguish between them in accordance with the following definition.

Definition 3-1   *All the digits of a number representing an approximate value except zeros which serve only to fix the position of the decimal point are called* significant digits *or* significant figures.

The following numbers have their significant digits underscored:

0.00<u>251</u>   <u>308</u>   <u>2.50</u>   <u>500.0</u>   <u>23.00</u>

The zero in 2.50 stands for a known digit and is significant; if the digit for that place were unknown, the number would be written as 2.5. The approximation 2.50 stands for a value between 2.495 and 2.505. Likewise, the zeros in the last numbers are significant.

The significant digits in decimal notation are apparent from the number itself except in one case. An integer ending with one or more zeros does not reveal the nature of the zero or zeros. In the so-called *scientific notation*, however, all significant digits are evident. In this notation a number is expressed as a product. The first factor has the significant digits with a decimal point just after the first digit, and the second factor is an integral power of 10. Employing scientific notation, for example, we can indicate the nature of the zeros in 800. To indicate that neither zero, one zero, or both zeros are significant, we express the number, respectively, as

$$8 \times 10^2 \qquad 8.0 \times 10^2 \qquad 8.00 \times 10^2$$

In each form, the significant digits appear in the first factor; and the product of the factors is 800.

A result obtained from calculations in which approximate values occur may have digits which are not significant. Such digits need to be discarded. The discarding of extra digits is called *rounding off*. In each case the new number is an approximation to the original number.

EXAMPLES   We round off 2.718 to three digits and have 2.72. Here we add 1 to the third digit of the given number in order to obtain the best three-figure approximation. For two-figure and one-figure approximations, we round off 2.718, respectively, to 2.7 and 3.

In rounding off 2.715 to three digits, we observe that 2.71 and 2.72 are equally valid. In all such cases, we shall arbitrarily *round off so that the last digit is an even number.* Accordingly, we choose 2.72.

If any discarded digits are to the left of the decimal point, their places are filled with zeros. Thus, when rounded off to two significant figures, 756.2 becomes 760, with the zero not significant, or, in scientific notation, 7.6×10^2.

We state here two rules for rounding off which are frequently employed and which are usually satisfactory:

Rule 1   *To find the sum (or difference) of two numbers representing approximations, write one number under the other and add (or subtract) in the usual way. Then round off the result so that the last digit retained is in the column farthest to the right in which both given numbers have significant digits.*

Rule 2   *To find the product (or quotient) of two numbers representing approxi-mations, multiply (or divide) in the usual way. Then round off the result to the smaller number of significant digits found in either of the given numbers.*

It is sometimes advantageous to do certain rounding off before multiplying or dividing numbers. If the significant digits in one number exceed those in the other number by two or more, we suggest that the first number be rounded off so that its excess in significant digits is only one. For example, to find the product of 3.1729 and 2.16, we could replace 3.1729 by 3.173 and then multiply.

EXAMPLE 1   Find the sum and difference of the numbers 84.037 and 2.72.

Solution

84.037		84.037	
2.72		2.72	
86.757	sum	81.317	difference

The column farthest to the right in which both given numbers have significant digits corresponds to the fourth digit in each of these results. We round off accordingly and have the approximate values 86.76 and 81.32.

EXAMPLE 2   Find the product of 0.0241 and 0.52.

Solution   The first number has three significant figures and the second two significant figures. Hence, their product should show two significant figures. Multiplying, we get

$$0.0241 \times 0.52 = 0.012532$$

Rounding off this value to two significant figures, we have 0.013, or, in scientific notation, 1.3×10^{-2}, in the final result.

Exercise 3-1

Consider the following numbers as approximate measurements and give in each case the values between which the correct measurement lies.

1. 506 feet   **2.** 28.3 pounds   **3.** 44.0 feet
4. 0.37 inch   **5.** 7.160 centimeters **6.** 43.40 pounds

Round off each number to (*a*) three significant digits, (*b*) two significant digits, (*c*) one significant digit.

7. 6444	**8.** 1624	**9.** 3060	**10.** 0.2056
11. 0.9992	**12.** 62.30	**13.** 274,351	**14.** 140.002482

15. Express 4000 in scientific notation, indicating that it has (*a*) three significant digits, (*b*) two significant digits, (*c*) one significant digit.

16. Express 2500 in scientific notation, indicating that it has (*a*) three significant digits, (*b*) two significant digits, (*c*) one significant digit.

Perform the indicated operations and give the result with the proper number of significant figures.

17. $3.8 + 2.05$	**18.** $2.06 + 8.5$	**19.** $9.60 - 7.65$
20. 2.6×3.4	**21.** 162×2.0	**22.** 11.0×14.0
23. $(4.60)(5.341)$	**24.** $(230)(0.33)$	**25.** $(612)(0.44)$
26. $68.3 \div 5.6$	**27.** $4.31 \div 0.13$	**28.** $(4.00)(3.0)$

3-2 Tables of trigonometric functions

The values of the trigonometric functions of angles are required in many problems and investigations. Although the exact values for certain special angles can be found, the exact values for an arbitrary angle cannot be determined. By methods of calculus, however, the values of the functions for any angle can be computed to any desired degree of accuracy. Table 2 in the Appendix gives the values of the functions of angles, at 10′ intervals, from 0° to 90°. This table is called *a table of natural trigonometric functions* to distinguish it from a table of the logarithms of the values of the functions, to be discussed later. The values of the functions, with a few exceptions, are unending decimals, and hence the tabular values are approximations. The table is called *a four-place table,* since the values of the functions are rounded off to four significant figures.

Angles from 0° to 45° are in the first column of the pages, and the next six columns contain the function values, with a particular function indicated at the top of each column. The function values of each angle are to the right along the line of the angle. Angles from 45° to 90° are in the last column of the pages. The values of the functions of an angle in this rightmost column are to the left along the line of the angle, with the functions indicated at the bottom of the page.

The table may be used to find a function value of a given angle or, inversely, to find the angle corresponding to a given function value. We shall explain the use of the table with some illustrative examples.

EXAMPLE 1   Find the sine, cosine, and tangent of 41° 20′.

Solution   We locate $41° 20'$ in the first column of Table 2. Then by looking across the table in line with the angle and reading from the columns headed by Sin, Cos, and Tan, we find

$\sin 41° 20' = 0.6604$
$\cos 41° 20' = 0.7509$
$\tan 41° 20' = 0.8796$

EXAMPLE 2   Find the sine, cosine, and cotangent of $73° 50'$.

Solution   We locate $73° 50'$ in the last column of Table 2, and read the function values which are in line with the angle and in the columns with Sin, Cos, and Cot at the bottom. Thus, we obtain

$\sin 73° 50' = 0.9605$
$\cos 73° 50' = 0.2784$
$\cot 73° 50' = 0.2899$

EXAMPLE 3   Given $\sin \theta = 0.8526$, find θ.

Solution   We look for the entry .8526 in a column headed by Sin or in a column with Sin at its foot. To locate the entry readily, we observe that from $0°$ to $90°$ the sine increases from 0 to 1. The number .8526 appears in Table 2 in the column with Sin at the bottom, and the angle must be read at the right. We find $\theta = 58° 30'$.

Exercise 3-2

Use Table 2 to find the value of each quantity.

1. $\sin 24° 40'$	**2.** $\tan 40° 30'$	**3.** $\cot 23° 00'$
4. $\cos 60° 20'$	**5.** $\sec 28° 10'$	**6.** $\csc 46° 50'$
7. $\cot 61° 00'$	**8.** $\sec 44° 30'$	**9.** $\csc 3° 10'$
10. $\cos 3° 40'$	**11.** $\sin 84° 00'$	**12.** $\sin 60° 10'$

Find θ in each of the following equations.

13. $\sin \theta = 0.1392$	**14.** $\cos \theta = 0.9981$
15. $\cot \theta = 7.953$	**16.** $\sec \theta = 1.480$
17. $\cos \theta = 0.7844$	**18.** $\sin \theta = 0.5831$
19. $\tan \theta = 0.5969$	**20.** $\csc \theta = 2.142$
21. $\sin \theta = 0.7153$	**22.** $\cos \theta = 0.5688$
23. $\cot \theta = 0.4699$	**24.** $\sec \theta = 2.295$
25. $\tan \theta = 1.393$	**26.** $\cot \theta = 0.8195$

3-3 Interpolation

In the preceding section, we learned how to find the functions of an angle which appears in the table and also how to find the angle corresponding to a function value. When the given angle is not listed in the table, we can find approximate values of its functions by a process called *interpolation*. According to this process, a function of an angle is proportional to the angle for small changes in the angle. Although this assumption is not entirely true, improved accuracy may be obtained by interpolation. Interpolation may also be used to find an angle whose given function value is not listed. We shall explain the procedure with some examples.

EXAMPLE 1   Find the value of sin 57° 14′.

Solution   The given angle, not listed in Table 2, has a value between the consecutive entries 57° 10′ and 57° 20′. Since 57° 14′ is four-tenths of the way from 57° 10′ to 57° 20′, we assume that sin 57° 14′ is also four-tenths of the way from sin 57° 10′ to sin 57° 20′. Reading directly from Table 2, we find sin 57° 10′ = 0.8403 and sin 57° 20′ = 0.8418. From 8403 to 8418 (neglecting decimal points for the moment) is 15 units, and $\frac{4}{10} \times 15 = 6$. Adding 6 to 8403, we have 8409; hence sin 57° 14′ = 0.8409.

The data and computations may be put in the following tabular form:

$$
10'\left[4'\begin{bmatrix} \sin 57^\circ\ 10' = 0.8403 \\ \sin 57^\circ\ 14' = \qquad ? \\ \sin 57^\circ\ 20' = 0.8418 \end{bmatrix}15 \right. \qquad
\begin{matrix} \tfrac{4}{10} \times 15 = 6 \\ 8403 + 6 = 8409 \\ \sin 57^\circ\ 14' = 0.8409 \end{matrix}
$$

Note: The difference between two consecutive entries of the table is called the *tabular difference*. In the preceding example the tabular difference of sin 57° 10′ and sin 57° 20′ is 0.0015. The part of the tabular difference used in the interpolation process is called the *correction*. In computing the correction, we may omit the decimal points until we write the final result. Usually it is necessary to round off in finding the correction. Had the preceding example required sin 57° 15′, we would have had, for the correction, $\frac{5}{10} \times 15 = 7.5$. We would round off this number to 8 in accordance with our agreement in Sec. 3-1. In no case should more digits be kept in the interpolated result than are found in the table.

EXAMPLE 2   Find the value of cos 37° 27′.

Solution   As we can see from the table, the values of the cosine decrease as the angle increases from 0° to 90°. Hence, cos 37° 27′ is less than cos 37° 20′ and greater than cos 37° 30′. Accordingly, the correction in this case is to be subtracted from the value of cos 37° 20′. We find cos 37° 20′ = 0.7951 and cos 37° 30′ = 0.7934. The tabular difference is 7951 − 7934 = 17 (neglecting decimal points). Since the angle 37° 27′

is $\frac{7}{10}$ of the way from 37° 20′ to 37° 30′, we take $\frac{7}{10}$ of 17 (equals 12, rounded off) as the correction. Subtracting this correction, we have 7951 − 12 = 7939; hence, cos 37° 27′ = 0.7939.

The quantities involved in the preceding discussion are as follows:

$$10' \left[7' \begin{bmatrix} \cos 37° \ 20' = 0.7951 \\ \cos 37° \ 27' = \qquad ? \\ \cos 37° \ 30' = 0.7934 \end{bmatrix} 17 \right. \qquad \begin{array}{l} \frac{7}{10} \times 17 = 12 \qquad \text{rounded off} \\ 7951 - 12 = 7939 \\ \cos 37° \ 27' = 0.7939 \end{array}$$

EXAMPLE 3   Find θ if cot θ = 0.5782.

Solution   We look for .5782 in a column with Cot at the top or Cot at the bottom. Our number is not in any of these columns, but its value is between two consecutive entries with Cot at the bottom of Table 2. Since the name of the function is at the bottom, we read the corresponding angles at the right. The data and computations may be arranged in tabular form as follows:

$$10' \left[\begin{bmatrix} \cot 59° \ 50' = 0.5812 \\ \qquad ? = 0.5782 \\ \cot 60° \ 00' = 0.5774 \end{bmatrix} 30 \right] 38 \qquad \begin{array}{l} \frac{30}{38} \times 10' = 8' \qquad \text{nearest integer} \\ 59° \ 50' + 8' = 59° \ 58' \\ \text{Hence } \theta = 59° \ 58' \end{array}$$

Note: It is advisable when first interpolating to write the quantities in detail, similar to that of the preceding problems. When the procedure is fully understood, the computations in most cases can be done mentally, and only the final result needs to be written. Also, it may be observed from the table that the sine, tangent, and secant increase as the angle increases from 0° to 90° and the cosine, cotangent, and cosecant decrease.

Exercise 3-3

Interpolate to find the function values.

1. sin 18° 25′   2. sin 33° 23′   3. tan 22° 35′
4. tan 12° 21′   5. sin 59° 37′   6. tan 66° 8′
7. cos 18° 47′   8. cos 31° 35′   9. cot 27° 34′
10. cot 23° 29′   11. cos 55° 23′   12. cos 81° 28′
13. sec 14° 51′   14. sec 47° 27′   15. csc 44° 26′
16. csc 49° 11′   17. sin 89° 21′   18. cos 4° 41′
19. cot 61° 9′   20. sec 72° 5′   21. csc 19° 18′

Interpolate to find each angle θ to the nearest minute.

22. sin θ = 0.6060   23. cos θ = 0.5070
24. sin θ = 0.7009   25. sin θ = 0.8991

26. $\tan \theta = 0.2477$	**27.** $\tan \theta = 0.7008$
28. $\cos \theta = 0.8834$	**29.** $\cos \theta = 0.0825$
30. $\cos \theta = 0.6600$	**31.** $\cos \theta = 0.3684$
32. $\cot \theta = 5.342$	**33.** $\cot \theta = 0.0123$
34. $\sec \theta = 1.052$	**35.** $\sec \theta = 1.482$
36. $\csc \theta = 3.906$	**37.** $\csc \theta = 1.500$
38. $\sin \theta = 0.9991$	**39.** $\cos \theta = 0.0931$
40. $\tan \theta = 0.0912$	**41.** $\cot \theta = 10.58$
42. $\tan \theta = 1.242$	**43.** $\tan \theta = 2.116$

3-4   Reduction to positive acute angles *

Having used a table to find the values of trigonometric functions of angles in the interval $0°$ to $90°$, we next consider the problem of finding the values of functions of angles outside this interval. Any angle greater than $360°$ or any negative angle in standard position is coterminal with a positive angle not exceeding $360°$. Hence the values of the functions of any angle are equal to those of some angle in the interval $0°$ to $360°$. We shall next show that the values of functions of angles in the interval $0°$ to $360°$, and therefore all angles, may be expressed in terms of values of functions of angles in the interval $0°$ to $90°$. We shall not include quadrantal angles in this discussion, however, since we have already obtained their function values (Sec. 2-4). We begin by introducing the idea of a reference angle.

Definition 3-2   *The positive acute angle which the terminal side of an angle in standard position makes with the x axis is called the refer- ence angle of the given angle.*

The terminal side of $213°$, for example, is in the third quadrant and makes an angle of $33°$ with the x axis. Hence $33°$ is the reference angle of $213°$, as exhibited in Fig. 3-1. Similarly, the reference angle of $-150°$ is $30°$, and the reference angle of $420°$ is $60°$.

If θ is a positive acute angle, then $(180° - \theta), (180° + \theta)$, and $(360° - \theta)$ are angles in the second, third, and fourth quadrants, respectively. In each case θ is the angle which the terminal side makes with the x axis and con- sequently is the reference angle of the original angle. The trigonometric functions of these angles, as we shall show, are expressible as functions of the reference angle θ.

Each of the angles $(180° - \theta)$ and θ is constructed in standard position in Fig. 3-2. The points P and P' on the terminal side are chosen so that

---

*An angle whose measure is between $0°$ and $90°$ or between $0°$ and $-90°$ is called an *acute angle*.

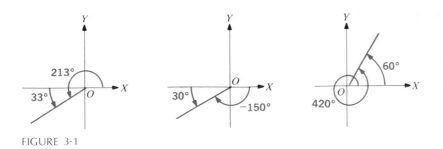

FIGURE 3-1

$OP = OP' = 1$. By the definitions of the sine and cosine, we have

$$\sin (180° - \theta) = \frac{MP}{OP} = \frac{MP}{1} = y$$

$$\cos (180° - \theta) = \frac{OM}{OP} = \frac{OM}{1} = x$$

$$\sin \theta = y' \quad \text{and} \quad \cos \theta = x'$$

We next observe that the right triangles OMP and $OM'P'$ are congruent, and consequently their corresponding sides have equal lengths. Since coordinates of points are directed distances, x and x', though of the same length, are not algebraically equal; each is the negative of the other. Hence $x = -x'$ and $y = y'$. And, recalling that $\tan \theta = \sin \theta / \cos \theta$, we may write

$$\sin (180° - \theta) = \sin \theta$$
$$\cos (180° - \theta) = -\cos \theta$$
$$\tan (180° - \theta) = -\tan \theta$$

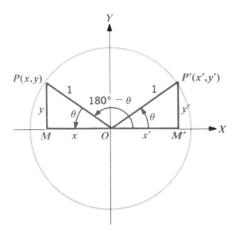

FIGURE 3-2

Referring to Fig. 3-3, where $x = -x'$ and $y = -y'$, we have

$$\sin(180° + \theta) = -\sin\theta$$
$$\cos(180° + \theta) = -\cos\theta$$
$$\tan(180° + \theta) = \tan\theta$$

Finally, in Fig. 3-4, $x = x'$ and $y = -y'$, and therefore

$$\sin(360° - \theta) = -\sin\theta$$
$$\cos(360° - \theta) = \cos\theta$$
$$\tan(360° - \theta) = -\tan\theta$$

The preceding formulas are called *reduction formulas*. We note that similar reduction formulas could be written for the cotangent, secant, and cosecant functions. We see then that the functions of the second-, third-, and fourth-quadrant angles are expressible as functions of angles between 0° and 90°. It may also be observed that each function value is equal to the same function value of the reference angle or is equal to its negative. Hence we have the following theorem.

Theorem 3-1   *Depending on the position of the terminal side, a trigonometric function value of an angle in any quadrant is either equal to the same function value of the reference angle or is equal to the negative of the same function value.*

EXAMPLE 1   Express $\sin 148°$ and $\tan 148°$ as functions of the reference angle.

Solution   The terminal side of 148° is in the second quadrant; this makes the sine positive and the tangent negative. The terminal side forms a 32°

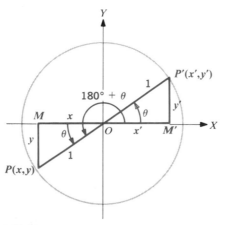

FIGURE 3-3

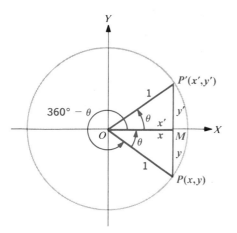

FIGURE 3-4

angle with the negative x axis. Hence the reference angle is $32°$, and we have

$$\sin 148° = \sin 32° \qquad \text{and} \qquad \tan 148° = -\tan 32°$$

EXAMPLE 2   Express $\cos 220°$ and $\cot 220°$ as functions of the reference angle.

Solution   The terminal side of $220°$ is in the third quadrant and the reference angle is $40°$. For a third-quadrant angle the cosine is negative and the cotangent is positive. Hence,

$$\cos 220° = -\cos 40° \qquad \text{and} \qquad \cot 220° = \cot 40°$$

The sign and the reference angle should be verified in each of the equations

$$\sin \frac{5\pi}{3} = -\sin \frac{\pi}{3}$$

$$\tan(-50°) = -\tan 50°$$

Exercise 3-4

Draw each angle in standard position and find the reference angle.

1. $160°$
2. $242°$
3. $320°$
4. $460°$
5. $-322°$
6. $-236°$
7. $-134°$
8. $-66°$
9. $\dfrac{3\pi}{5}$
10. $\dfrac{7\pi}{6}$
11. $\dfrac{8\pi}{5}$
12. $\dfrac{9\pi}{4}$

Give the value of each of the following quantities. Use Table 2.

13. $\sin 153°$ **14.** $\cos 194°$ **15.** $\tan (-63°)$

16. $\cot 284°$ **17.** $\sec 213°$ **18.** $\sin 92° \; 20'$

19. $\cos 121° \; 20'$ **20.** $\sin (-148°)$ **21.** $\csc (-141°)$

22. $\tan 800°$ **23.** $\cos 1200°$ **24.** $\sec (-790°)$

3-5 The right triangle

The right triangle is a figure of special importance; it occurs in many ordinary experiences and has numerous applications in engineering, physics, and other sciences. Hence, the right triangle is given special attention in trigonometry courses. We shall begin our study by learning how to find the values of the trigonometric functions of the acute angles from the lengths of the sides of the triangle.

Figure 3-5 shows a right triangle with the acute angle A in standard position. The coordinates of the vertex of angle B are $x = b, y = a$, and the distance $r = c$. We observe that a is the length of the side opposite A, b is the length of the side adjacent to A, and c is the length of the hypotenuse. Then using these terms and applying the definitions of the trigonometric functions (Sec. 2-3), we have

$$\sin A = \frac{\text{side opposite}}{\text{hypotenuse}} \qquad \csc A = \frac{\text{hypotenuse}}{\text{side opposite}}$$

$$\cos A = \frac{\text{side adjacent}}{\text{hypotenuse}} \qquad \sec A = \frac{\text{hypotenuse}}{\text{side adjacent}}$$

$$\tan A = \frac{\text{side opposite}}{\text{side adjacent}} \qquad \cot A = \frac{\text{side adjacent}}{\text{side opposite}}$$

The functions of A depend only on the lengths of the sides of the triangle; they are independent of a coordinate system. Henceforth, then, we may obtain the values of the functions of either acute angle of a right triangle

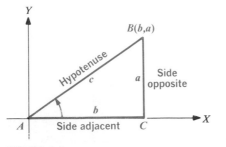

FIGURE 3-5

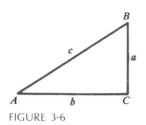

FIGURE 3-6

by picking out the opposite side and the adjacent side and applying the above results. It is essential to remember, however, that this method applies only to the acute angles of a *right* triangle.

In dealing with right triangles, we shall usually let A and B stand for the acute angles and C for the right angle. Also, we shall let a, b, and c stand for the lengths of the sides opposite A, B, and C, respectively.

If the sum of two positive angles is $90°$, the angles are said to be *complementary*. It follows from this definition that the acute angles of a right triangle are complementary. We shall use this fact to see how the functions of complementary angles are related. Referring to Fig. 3-6, we observe that side a is opposite angle A and adjacent to angle B, and that side b is opposite B and adjacent to A. Hence, we may write

$$\sin A = \frac{a}{c} = \cos B \qquad \csc A = \frac{c}{a} = \sec B$$

$$\cos A = \frac{b}{c} = \sin B \qquad \sec A = \frac{c}{b} = \csc B$$

$$\tan A = \frac{a}{b} = \cot B \qquad \cot A = \frac{b}{a} = \tan B$$

In these equations the functions appear in the pairs *sine* and *cosine, tangent* and *cotangent, secant* and *cosecant.* The functions in each of these pairs are called *cofunctions.* For example, the cosine is the cofunction of the sine, and the sine is the cofunction of the cosine. The equations show, then, that a function of A is equal to its cofunction of B. Since any two complementary angles may be the acute angles of a right triangle, we have established the following theorem:

Theorem 3-2   *Any trigonometric function of a positive acute angle is equal to the corresponding cofunction of the complementary angle.*

Conversely, if a function of a positive acute angle is equal to the cofunction of another positive acute angle, then the angles are complementary. The student may prove this statement.

EXAMPLE 1   Write each of the quantities $\sin 40°$, $\cot 70°$, $\sec 60°$ as a function of the complementary angle.

Solution   We subtract the angle in each case from $90°$ to obtain its complement, then we replace each angle by its complement and each function by its cofunction. Hence,

$$\sin 40° = \cos 50°$$
$$\cot 70° = \tan 20°$$
$$\sec 60° = \csc 30°$$

EXAMPLE 2   If $\tan 3\theta = \cot (\theta - 6°)$ and if 3θ and $(\theta - 6°)$ are positive acute angles, then find θ.

Solution   Since the tangent and cotangent are cofunctions, the angles are complementary. Therefore, we have

$$3\theta + (\theta - 6°) = 90°$$
$$4\theta = 96°$$
$$\theta = 24°$$

Exercise 3-5

1. Write the values of the trigonometric functions of angles A and B in each right triangle of Fig. 3-7.

Draw a right triangle and write the values of the trigonometric functions of angles A and B in each of the following problems.

2. $a = 5, b = 12$ 3. $a = 24, c = 25$ 4. $a = 35, c = 37$
5. $a = 3, b = 7$ 6. $b = 1, c = 4$ 7. $a = 7, b = 8$

Express each of the following quantities as a function of the complementary angle.

8. $\sin 35°$ 9. $\cos 65°$ 10. $\tan 68°$ 11. $\cot 51°$

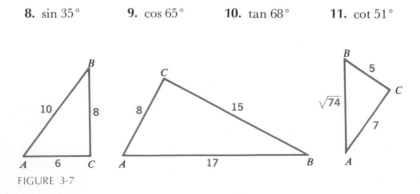

FIGURE 3-7

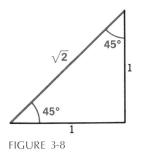

FIGURE 3-8

12. $\csc\dfrac{\pi}{4}$ **13.** $\sec\dfrac{3\pi}{10}$ **14.** $\sin\dfrac{\pi}{3}$ **15.** $\cos\dfrac{2\pi}{5}$

Find θ if functions of positive acute angles are involved in each equation.

16. $\cot\theta = \tan 70°$ **17.** $\sin\tfrac{1}{2}\theta = \cos\tfrac{1}{2}\theta$

18. $\cos 3\theta = \sin 2\theta$ **19.** $\tan(\theta + 44°) = \cot\theta$

20. $\sin\left(\theta + \dfrac{\pi}{3}\right) = \cos\left(\theta - \dfrac{\pi}{8}\right)$ **21.** $\csc\left(\theta + \dfrac{\pi}{5}\right) = \sec\left(\theta + \dfrac{\pi}{10}\right)$

22. $\cot\left(3\theta + \dfrac{\pi}{9}\right) = \tan\left(\theta + \dfrac{\pi}{4}\right)$ **23.** $\sec\left(\theta + \dfrac{\pi}{7}\right) = \csc\left(2\theta + \dfrac{2\pi}{7}\right)$

24. If A and B are positive acute angles and $\sin A = \cos B$, prove that $A + B = 90°$.

3-6 Functions of 30°, 45°, and 60°

There are certain special angles whose function values can be determined geometrically. Among these are the angles 30°, 45°, and 60°.

The functions of 45° may be read directly from a right triangle whose acute angles are each equal to 45°. For convenience, we let the length of the equal sides be 1, and therefore the hypotenuse is equal to $\sqrt{2}$. Then, from Fig. 3-8, we write

$$\sin 45° = \frac{1}{\sqrt{2}} \qquad \cos 45° = \frac{1}{\sqrt{2}} \qquad \tan 45° = 1$$
$$\csc 45° = \sqrt{2} \qquad \sec 45° = \sqrt{2} \qquad \cot 45° = 1$$

To obtain the functions of 30° and 60°, we use one of the right triangles formed by bisecting an angle of an equilateral triangle. It is evident that the right triangle thus constructed (Fig. 3-9) has acute angles of 30° and 60° and also that the hypotenuse is double the side opposite the 30° angle. If we select the convenient lengths 2 and 1 for the hypotenuse and the shorter

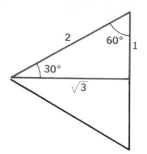

FIGURE 3-9

side, then the length of the remaining side is $\sqrt{3}$. From the triangle, we have

$$\sin 30° = \frac{1}{2} \qquad \cos 30° = \frac{\sqrt{3}}{2} \qquad \tan 30° = \frac{1}{\sqrt{3}}$$

$$\csc 30° = 2 \qquad \sec 30° = \frac{2}{\sqrt{3}} \qquad \cot 30° = \sqrt{3}$$

$$\sin 60° = \frac{\sqrt{3}}{2} \qquad \cos 60° = \frac{1}{2} \qquad \tan 60° = \sqrt{3}$$

$$\csc 60° = \frac{2}{\sqrt{3}} \qquad \sec 60° = 2 \qquad \cot 60° = \frac{1}{\sqrt{3}}$$

It is advantageous to be able to write readily the values of the functions of 30°, 45°, and 60° since these values are used so frequently. A better plan than memorizing the values, however, is to visualize the right triangles of Figs. 3-8 and 3-9 and then read the values from the mental picture.

The methods of evaluating the functions of 30°, 45°, and 60° will also serve to yield the functions of angles in any quadrant if the terminal side makes one of these angles with the positive or negative x axis. This is done by using the appropriate reduction formulas (Sec. 3-4) or, alternatively, from a figure, as illustrated in the following example:

EXAMPLE   Construct the angle 330° in standard position and find the sine, cosine, and tangent of the angle.

Solution   The terminal side of 330° is in the fourth quadrant and makes an angle of 30° with the positive x axis (Fig. 3-10). We select a point P on the terminal side 2 units from the origin and draw MP, thus forming the reference triangle OMP. The coordinates of P are $(\sqrt{3}, -1)$. Then, applying the definitions of the trigonometric functions, we obtain

$$\sin 330° = -\frac{1}{2} \qquad \cos 330° = \frac{\sqrt{3}}{2} \qquad \tan 330° = -\frac{1}{\sqrt{3}}$$

Exercise 3-6

Substitute the values of the trigonometric functions and show that each equation 1 through 12 is true.

1. $\sin^2 45° + \cos^2 45° = 1$†
2. $\cos 60° = 2 \sin 30° \cos 30°$
3. $\sec^2 30° - 1 = \tan^2 30°$
4. $\csc^2 30° - 1 = \cot^2 30°$
5. $\cos 60° = 2 \cos^2 30° - 1$
6. $\cos^2 30° = \sin^2 30° + \cos 60°$
7. $2 \sin^2 30° + \cos 60° = 1$
8. $1 - \sin^2 60° = \cos^2 60°$

9. $\dfrac{2 \tan 30°}{1 - \tan^2 30} = \tan 60°$
10. $\dfrac{\tan 60° - \tan 30°}{1 + \tan 60° \tan 30°} = \tan 30°$

11. $\dfrac{\cot^2 30° - 1}{2 \cot 30°} = \cot 60°$
12. $\dfrac{1 + \cos 60°}{\sin 60°} = \cot 30°$

Evaluate each of the following expressions.

13. $\dfrac{1 + \cot 30° \cot 45°}{\cot 30° - \cot 45°}$
14. $\dfrac{\cos 30° \sin 30°}{\cos 60° - \sin 60°}$

15. $\sin 60° \cos 45° + \cos 60° \sin 45°$
16. $\cos 45° \cos 60° + \sin 45° \sin 60°$
17. $\sin 45° \cos 30° - \cos 45° \sin 30°$

Draw each angle in standard position, complete a reference triangle, and find the values of the angle's sine, cosine, and tangent.

18. $120°$ 19. $135°$ 20. $225°$ 21. $210°$
22. $240°$ 23. $315°$ 24. $-60°$ 25. $-210°$

†The notation $\sin^2 \theta$, read "sine squared theta," means $\sin \theta \cdot \sin \theta$ or $(\sin \theta)^2$, and similarly for the other functions.

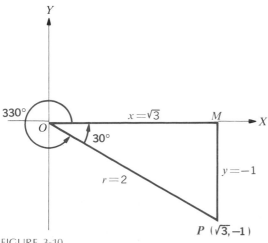

FIGURE 3-10

3-7   Solution of right triangles

The three sides and the three angles of a triangle are called its *parts*. The process of finding the unknown parts from the given parts is called *solving the triangle*. In this section we shall consider the problem of solving right triangles. It is possible to find the remaining parts of a right triangle if, in addition to the right angle, one side and any other part (side or angle) are known. The key principle in solving a right triangle is the fact that a trigonometric function of an acute angle yields an equation involving three quantities, namely, two sides and the value of the function formed by the ratio of the sides. In solving triangles, we shall regard the given parts as approximate quantities. This means that the accuracy of the computed parts depends on the accuracy of the given parts. Accordingly, a computed side should be rounded off to as many significant figures as contained in the given side or sides (Sec. 3-1). Since angles are also involved, we need to know what accuracy in angles is comparable with a given accuracy in the sides. The corresponding accuracies indicated in the table, and which we shall use, are approximately correct.

Sides	Angles
Two significant figures	Nearest degree
Three significant figures	Nearest multiple of 10′
Four significant figures	Nearest minute

A computed angle, by this plan, should be expressed to the nearest degree, nearest multiple of 10′, or nearest minute according as the sides have two, three, or four significant figures. It is a good plan, however, when the sides have fewer significant figures than the table, to carry along in the computing process one step greater accuracy than that of the given data and finally to round off properly the values of the computed parts. This is illustrated in the following example.

EXAMPLE 1   A right triangle has $a = 38$ and $B = 61°$. Find A, b, and c.

Solution   We first draw a right triangle approximately to scale and mark the known parts (Fig. 3-11). The acute angles are complementary, and we have at once

$$A = 90° - 61° = 29°$$

To find b, we write, from the definition of tangent,

$$\frac{b}{38} = \tan 61°$$

$$b = 38 \tan 61°$$

Hence, using Table 2, we find

$b = 38(1.80)$
$= 68.4$
$= 68$ rounded off

Secant B may conveniently be used in finding c. Thus,

$\dfrac{c}{38} = \sec 61°$

$c = 38 \sec 61°$
$= 38(2.06)$
$= 78$ rounded off

From the given parts, we have obtained $A = 29°$, $b = 68$, and $c = 78$. We rounded off the sides to two significant figures because the given side has two significant figures. Also note that we shortened the computations by rounding off the values of $\tan 61°$ and $\sec 61°$ to three significant figures—one more than the number of significant figures in the given sides.

EXAMPLE 2 Solve the right triangle which has $a = 61.84$ and $c = 85.76$.

Solution From the right triangle (Fig. 3-12), we write

$\sin A = \dfrac{61.84}{85.76} = 0.7211$

Using Table 2 and interpolating, we find

$A = 46° 9'$ and $B = 90° - 46° 9' = 43° 51'$

From the definition of cosine, we now obtain

$\dfrac{b}{85.76} = \cos 46° 9'$

$b = 85.76 \cos 46° 9' = 85.76(0.6928)$
$= 59.41$ rounded off

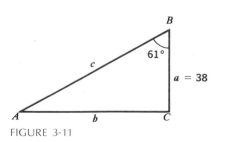

FIGURE 3-11

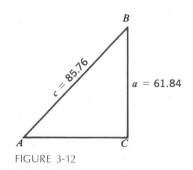

FIGURE 3-12

The results are $A = 46° 9'$, $B = 43° 51'$, and $b = 59.41$.

Although the preceding examples show how to solve right triangles, we list the following directions:

1. Use the given parts and construct a right triangle approximately to scale. Mark the known parts in the figure and indicate the unknown parts with letters. The figure, exhibiting roughly the values of the parts to be computed, may serve as a check against large errors.
2. For each unknown part choose a formula which will yield that part. It is better, where possible, to find a part from given parts rather than from computed parts. This plan tends to give more accurate results and avoids carrying forward an erroneous computation.
3. Arrange the work in a neat, orderly way.

As a further remark, we remind the student that two different methods of computing a part of a triangle will sometimes yield slightly different results. Accordingly, a result obtained correctly may sometimes differ from the answer given in the book.

Exercise 3-7

Find the remaining parts of each right triangle. Give the results properly rounded off.

1. $a = 22$, $A = 35°$	**2.** $b = 33$, $B = 40°$
3. $c = 6.5$, $B = 28°$	**4.** $c = 30$, $A = 26°$
5. $b = 42$, $A = 53°$	**6.** $a = 16$, $B = 46°$
7. $b = 60$, $a = 82$	**8.** $a = 5.0$, $b = 4.6$
9. $a = 6.1$, $c = 8.4$	**10.** $a = 0.62$, $b = 0.46$
11. $a = 45$, $b = 83$	**12.** $b = 3.4$, $c = 5.5$
13. $a = 251$, $A = 35° 20'$	**14.** $b = 210$, $B = 55° 40'$
15. $b = 0.522$, $A = 52° 20'$	**16.** $a = 23.1$, $A = 43° 30'$
17. $a = 3.50$, $b = 4.20$	**18.** $a = 0.251$, $b = 0.350$
19. $a = 43.0$, $c = 53.0$	**20.** $b = 1.64$, $c = 2.60$

Solve the right triangles in problems 21 through 28. Express the computed sides to four significant figures and the angles to the nearest minute.

21. $c = 4000$, $A = 60° 20'$	**22.** $c = 3000$, $B = 50° 12'$
23. $a = 4100$, $b = 5200$	**24.** $a = 3.572$, $c = 5.000$
25. $a = 0.5763$, $B = 50° 30'$	**26.** $b = 5331$, $A = 54° 18'$
27. $a = 48.32$, $b = 34.21$	**28.** $a = 4.320$, $b = 5.324$

29. The length of the base of an isosceles triangle is 79.0 inches, and the angle opposite the base is 76° 00′. Find the other parts of the triangle.

30. Find the altitude of an equilateral triangle each of whose sides is 142 feet long.

31. Find the length of a side of an equilateral triangle which is inscribed in a circle of radius 10.4 feet.

32. Find the perimeter of a regular pentagon (five-sided polygon with equal sides and angles) which is inscribed in a circle with a radius of 24 inches.

33. A chord of length 16 inches is drawn in a circle with a radius of 24 inches. Find (*a*) the central angle subtended by the chord, (*b*) the distance from the center of the circle to the midpoint of the chord.

34. Find the perimeter of a regular octogon which is (*a*) inscribed in a circle of a radius 21 inches, (*b*) circumscribed about the circle.

3-8   Applications of right triangles

In certain problems of a practical nature, new terms arise. At this point we define some terms pertaining to angles.

Suppose *E* stands for the position of an observer's eye and *O* for a point on an object which he is viewing (Fig. 3-13). The line from *E* to *O* is called the *line of sight*. Let *EH* be a horizontal line in the vertical plane containing *EO*. Then, if *O* is above *E*, the angle *HEO* is called the *angle of elevation*. If *O* is lower than *E*, the angle *HEO* is called the *angle of depression*.

In plane surveying and marine navigation, the direction from one point to another is indicated in a special way. Let NS (Fig. 3-14) stand for a horizontal line in the north-south direction. Let *O* be a point on NS and *A* a point in the horizontal plane through NS. The acute angle which *OA* makes with NS is called the *bearing* of the point *A* from *O*. The angle is also the bearing of the line *OA*. The bearing of *A* may be precisely indicated by giving the size of the angle and specifying whether it is measured from *O*N or *O*S and whether to the east or west. This information is expressed

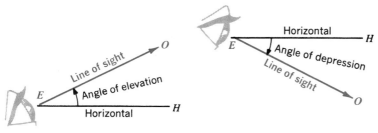

FIGURE 3-13

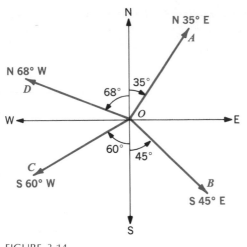

FIGURE 3-14

in abbreviated form. Thus, in the figure, OA is in a direction $35°$ east of north, and the bearing is written N $35°$ E. The bearings of some other points are also indicated.

EXAMPLE 1   From a point 124 feet from the foot of a tower and on the same level, the angle of elevation of the top of the tower is $36° 20'$ and the angle of elevation of the top of a water tank supported by the tower is $40° 50'$. Find (a) the height of the tower; (b) the depth of the tank.

Solution (a)   To find the height of the tower, we use the right triangle ABC (Fig. 3-15), where CB, or h, denotes the height of the tower. Thus,

$$\frac{h}{124} = \tan 36° 20'$$

$$h = 124 \tan 36° 20'$$
$$= 124(0.7355) = 91.20$$

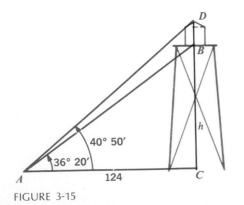

FIGURE 3-15

Solution (*b*)   Referring to the right triangle *ADC*, where $x = CD$ denotes the distance from the ground to the top of the tank, we obtain

$$x = 124 \tan 40° \ 50' = 124(0.8642) = 107.2$$

Then $BD = x - h = 107.2 - 91.20 = 16.0$. We now round off to three significant figures and express the height of the tower as 91.2 feet and the depth of the tank as 16.0 feet.

EXAMPLE 2   Two points, *A* and *B*, are 526 feet apart on a level stretch of road leading to a hill. The angle of elevation of the hilltop from *A* is 26° 30′, and the angle of elevation from *B* is 36° 40′. Find how high the hill extends above *AB* produced.

Solution   The given data are marked in Fig. 3-16. We let *h* stand for the height of the hill and *x* for the distance *BC*. From the right triangles *ADC* and *BDC*, we write

$$\cot 26° \ 30' = \frac{526 + x}{h}$$

and

$$\cot 36° \ 40' = \frac{x}{h}$$

These equations, when cleared of fractions, become

$$h \cot 26° \ 30' = 526 + x \qquad \text{and} \qquad h \cot 36° \ 40' = x$$

We observe that each equation contains two unknowns and that the two right triangles cannot be solved separately. Hence, we combine the equations to eliminate *x*. Thus we get

$$h \cot 26° \ 30' = 526 + h \cot 36° \ 40'$$
$$h \cot 26° \ 30' - h \cot 36° \ 40' = 526$$

$$h = \frac{526}{\cot 26° \ 30' - \cot 36° \ 40'}$$

$$= \frac{526}{2.006 - 1.343} = \frac{526}{0.663} = 793 \text{ feet}$$

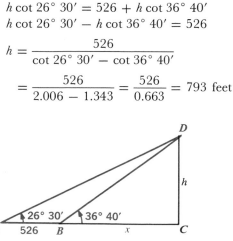

FIGURE 3-16

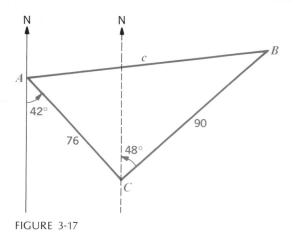

FIGURE 3-17

EXAMPLE 3   A ship sails a distance of 76 miles in the direction S 42° E and then sails a distance of 90 miles in the direction N 48° E. Find the distance and bearing of the ship's final position from its original position.

Solution   Referring to Fig. 3-17, we see that triangle ABC is a right triangle. So we first find angle A and then the hypotenuse c. Thus

$$\tan A = \frac{90}{76} = 1.184$$

$$A = 49° \; 50'$$

$$\frac{c}{90} = \csc 49° \; 50'$$

$$c = 90(1.309) = 117.8$$

We round off 49° 50′ to 50°. Then, since 180° − (42° + 50°) = 88°, we have N 88° E as the bearing of AB. Next, rounding off 117.8, we take 120 miles as the distance of the ship from its starting point.

Exercise 3-8

1. From a point 120 feet from the foot of a tower and on the same level, the angle of elevation of the top of the tower is 35° and the angle of elevation of the top of the water tank mounted on the tower is 42°. Find (a) the height of the tower; (b) the depth of the tank.
2. Two points, A and B, are 400 feet apart on a level stretch of road leading toward a hill. The angle of elevation of the hilltop from A is 25° and the angle of elevation from B is 42°. Find how high the hill extends above the road level.
3. A ship sails a distance of 70 miles in the direction N 38° E and then sails a

distance of 20 miles S 52° E. Find the distance and bearing of the ship's final position from its original position.

4. A surveyor observes the angle of elevation of the top of a vertical pole to be 27°. From a point in the same horizontal plane and 130 feet farther away he observes the angle of elevation to be 20°. If the two points of observation and the top of the pole are in the same vertical plane, find the height of the pole.

5. In order to find the distance between two points A and B at opposite ends of a lake, a surveyor laid off a base line BC at right angles to the line of sight AB and equal to 200 feet. Using a transit he found the angle BCA to equal 65°. Find the distance AB.

6. A ladder 14 feet long is leaning against a house. The foot of the ladder is 4.8 feet from the house. Find the angle of elevation of the ladder and the height it reaches on the house.

7. From a point 230 feet from the base of a smokestack, the angle of elevation of the top is 28° 30′. Find the height of the smokestack.

8. From a tower 70 feet high, the angle of depression of a car is 32°. How far is the car from the tower?

9. From a cliff 130 feet above the shore line, the angle of depression of a ship is 22° 40′. Find the distance from the ship to a point on the shore directly below the observer.

10. The shadow of a vertical pole is 24.8 feet long when the angle of elevation of the sun is 42° 00°. Find the height of the pole.

11. From a point on a lighthouse 108 feet above water, the angle of depression of a boat is 10° 10′. How far is the boat from the lighthouse?

12. What is the angle of elevation of the sun when a vertical pole 18 feet high casts a shadow 11 feet long?

13. In measuring the length of the east-west boundary of a piece of land, a surveyor comes to a marshy place. He diverts his path from a point P on the boundary line directly north to Q and from Q returns to R on the line. If the bearing of R from Q is S 60° 00′ E and the distance PQ is 320 feet, find the distance PR.

14. In measuring the length of the east-west boundary of a piece of land, a surveyor comes to an obstruction in his path. He then directs his path from a point P on the boundary line to a point Q and returns to R on the line. If the bearing of Q from P is S 50° E, the bearing of R from Q is N 40° E, and the distance $QR = 340$ feet, find the distance PR.

15. Two ships leave the same port. One travels 26 miles per hour in the direction N 40° E and the other travels 34 miles per hour in the direction S 50° E. Find the distance between the ships at the end of 2 hours.

16. A smokestack is 140 feet from a building. From a window of the building the angle of elevation of the top of the smokestack is 32° and the angle of depression of the base is 24°. Compute the height of the smokestack and the distance from the ground to the point of observation in the window.

17. From a point on a level with the base of a building and 420 feet away, the angle of elevation of the bottom of a flagpole on the building is 22° 20′ and the angle of elevation of the top of the pole is 23° 00′. Find the length of the pole.

18. From a point on level ground 250 feet from a church, the angle of elevation of the top of a spire on the building is 26° and the angle of elevation of the bottom of the spire is 24°. What is the length of the spire?

19. Two points A and B are 44 feet apart on a level stretch of road leading to a hill. The angle of elevation of the hilltop from A is 22° and the angle from B is 32°. Find how high the hill extends above the road level.

20. Generalize problem 19 by taking the distance from A to B as a feet and the angles of elevation from A and B, respectively, as θ and ϕ.

21. The angle of elevation of the top of a building from a point A, across the street and 22 feet above the ground, is 32°. From a point B, which is 13 feet vertically above A, the angle of elevation is 20°. How high is the building?

22. A balloon is in the vertical plane of two consecutive milestones of a road. The angles of depression of the stones as viewed from the balloon are 15° and 17°. Find the height of the balloon if (a) the stones are on opposite sides of the balloon, (b) the stones are on the same side of the balloon.

23. A balloon is in the vertical plane of two stones a feet apart. The angles of depression of the stones as viewed from the balloon are θ and ϕ. Find height of the balloon if (a) the stones are on opposite sides of the balloon, (b) the stones are on the same side of the balloon.

24. From a tower 124 feet high, the angles of depression of two rocks which are on a horizontal line through the base of the tower are 16° 00′ and 12° 00′. Find the distance between the rocks if they are on (a) opposite sides of the tower, (b) same side of the tower.

25. Find the distance between the rocks in problem 24 if one rock is directly south, and the other directly east, of the tower. The other data are unchanged.

26. A ship sailing in the direction S 42° W passes a point A directly east of a lighthouse. If the angle of elevation from A to the top of the lighthouse is 19°, find the angle of elevation when the ship is closest to the lighthouse.

27. A ship sailing in the direction N 54° E passes a point A directly south of a lighthouse. If the angle of elevation of the top of the lighthouse from A is 24°, find the angle of elevation when the ship is closest to the lighthouse.

Chapter 3 Review exercise

Interpolate to find the value of each of the following quantities.

1. tan 65° 9′   **2.** cos 71° 5′   **3.** sin 16° 17′

Interpolate to find the angle θ to the nearest minute.

4. $\cos \theta = 0.7483$ **5.** $\sin \theta = 0.9845$ **6.** $\tan \theta = 1.244$

Draw each of the following angles in standard position and find the reference angle.

7. $243°$ **8.** $235°$ **9.** $-317°$ **10.** $-42°$

Give the value of each of the following quantities. Use Table 2.

11. $\cos 141°$ **12.** $\sin 217°$ **13.** $\tan(-168°)$

Draw a right triangle ABC and write the values of $\sin A$, $\cos A$, and $\tan A$.

14. $a = 5, b = 12$ **15.** $a = 4, b = 6$ **16.** $b = 3, c = 7$

17. Construct the angle $240°$ in standard position and find the sine, cosine, and tangent of the angle.

18. A right triangle has $a = 40$ and $B = 63°$. Find A, b, and c.

19. Two ships leave the same port at the same time. One travels 20 miles per hour in the direction N $34°$ E and the other travels 30 miles per hour in the direction S $56°$ E. Find the distance between the ships at the end of 3 hours.

trigonometric identities

4-1 Conditional equations and identities

A statement that two quantities are equal is called an *equation*. The customary way of writing an equation is to place the symbol $=$ (read "is equal to") between the equal quantities. An equation then has two members, the left member and the right member. Equations usually have one or more letters which are regarded as *variables*. Numbers which, when substituted for the variables, make the two members of the equation equal are said to *satisfy* or be a *solution* of the equation. The totality of solutions is called the *solution set*.

EXAMPLES

$$x - 3 = 4 \tag{1}$$

$$x^2 - 2x - 15 = 0 \tag{2}$$

$$(x - 2)^2 = x^2 - 4x + 4 \tag{3}$$

$$\frac{1}{x - 1} - \frac{1}{x} = \frac{1}{x(x - 1)} \tag{4}$$

The first equation is satisfied when $x = 7$ and for no other number. The second equation is satisfied for $x = -3$ and $x = 5$; any other value for x makes the two members of the equation unequal. The right member of equation (3) is the result of the operation indicated on the left side. Clearly, then, the equation is satisfied when x is replaced by any real number. The right member of equation (4) is the sum of the fractions of the left member, and the equation is true for all values of x except $x = 0$ and $x = 1$. Either of these values produces a zero in a denominator on the left and in the denominator on the right. Consequently, we say that each member of the equation is *undefined* for these values of x.

The two types of equations which we have illustrated are named in accordance with the following definitions.

Definition 4-1   *An equation which is satisfied by some, but not all, of the values of the variables for which the members of the equation are defined is called a* conditional equation.

Definition 4-2   *An equation which is satisfied by all the values of the variables for which the members of the equation are defined is called an* identity.

According to these definitions, equations (1) and (2) are conditional equations, and equations (3) and (4) are identities. The solution sets of the four equations are respectively

$\{7\}$ $\{-3,5\}$ $\{x \,|\, x$ is a real number$\}$

$\{x \,|\, x$ is a real number not equal to 0 or 1$\}$

In this chapter we shall consider equations which involve trigonometric functions. An equation of this kind can be classed as a conditional trigonometric equation, a trigonometric identity, or an equation with no solution.

EXAMPLES

$$\csc \theta = \frac{1}{\sin \theta} \tag{5}$$

$$\tan \theta = \cot \theta \tag{6}$$

$$\sin \theta = 2 \tag{7}$$

The members of equation (5) are not defined for any angle which in standard position has its terminal side along the x axis. But the equation is satisfied for all other angles, and therefore is an identity. Equation (6) is satisfied for all angles in standard position whose terminal side bisects any of the four quadrants. Hence the equation is a conditional equation. The last equation has no solution because the sine function never exceeds 1 in absolute value.

Conditional trigonometric equations and trigonometric identities are of vast importance. Many practical problems require solutions of trigonometric equations; and aside from their employment in purely theoretical considerations, trigonometric identities are invaluable in a great diversity of problems, particularly in physics and the engineering sciences. We shall first study identities and in a later chapter take up conditional trigonometric equations.

4-2 The fundamental identities

We list here eight relations of the trigonometric functions. Each relation or equation is true for all angles for which the members of the equation are defined and therefore is an identity. These identities follow quite readily from the definitions of the trigonometric functions and are called *fundamental identities*. The identities are numbered and will later be referred to by the numbers assigned here.

In our discussion, we shall speak of the domain of θ as being a set of angles. The domain, however, may just as well be chosen as a set of real numbers (Sec. 2-5).

$$\sin \theta = \frac{1}{\csc \theta} \qquad \text{or} \qquad \csc \theta = \frac{1}{\sin \theta} \tag{8}$$

$$\cos \theta = \frac{1}{\sec \theta} \qquad \text{or} \qquad \sec \theta = \frac{1}{\cos \theta} \tag{9}$$

$$\tan \theta = \frac{1}{\cot \theta} \qquad \text{or} \qquad \cot \theta = \frac{1}{\tan \theta} \tag{10}$$

$$\tan \theta = \frac{\sin \theta}{\cos \theta} \tag{11}$$

$$\cot \theta = \frac{\cos \theta}{\sin \theta} \tag{12}$$

$$\sin^2 \theta + \cos^2 \theta = 1 \tag{13}$$

$$1 + \tan^2 \theta = \sec^2 \theta \tag{14}$$

$$1 + \cot^2 \theta = \csc^2 \theta \tag{15}$$

We have already obtained identities (8), (9), and (10) in Sec. 2-4. For convenience in deriving the remaining identities, we refer to Fig. 4-1, in which

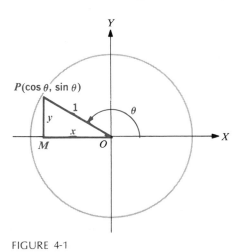

FIGURE 4-1

the angle θ is drawn in standard position, and the unit circle has its center at the origin (O). The terminal side of θ therefore cuts the circle at the point P, so that $r = 1$. Then, by definition,

$$\sin \theta = \frac{y}{1} = y \qquad \text{and} \qquad \cos \theta = \frac{x}{1} = x$$

Hence, the coordinates of P are $(\cos \theta, \sin \theta)$. From the definitions of $\tan \theta$ and $\cot \theta$, we can now write

$$\tan \theta = \frac{y}{x} = \frac{\sin \theta}{\cos \theta} \qquad \text{and} \qquad \cot \theta = \frac{x}{y} = \frac{\cos \theta}{\sin \theta}$$

Applying the Pythagorean theorem to the right triangle OMP in the figure, we have the equation $y^2 + x^2 = 1$, and since $y = \sin \theta$ and $x = \cos \theta$, we also have the identity

$$\sin^2 \theta + \cos^2 \theta = 1 \tag{13}$$

Next, we divide each member of this equation by $\cos^2 \theta$, and get

$$\frac{\sin^2 \theta}{\cos^2 \theta} + 1 = \frac{1}{\cos^2 \theta}$$

Identity (14) now follows because $\tan^2 \theta = \sin^2 \theta / \cos^2 \theta$ and $1/\cos^2 \theta = \sec^2 \theta$. Finally, we may establish identity (15) by dividing each member of (13) by $\sin^2 \theta$.

Identity (13) is true for all angles—positive, negative, and zero. The remaining fundamental identities do not hold for certain quadrantal angles. As we have already observed (Sec. 2-4), the tangent and secant functions are not defined if the terminal side of θ is along the y axis, and the cotangent and cosecant functions are not defined if the terminal side is along the x axis. This means that identities (9), (10), (11), and (14) are not valid if, in radian measure and with n any integer, $\theta = \frac{1}{2}\pi + n\pi$; and (8), (10), (12), and (15) are not valid if $\theta = n\pi$.

Because of their great importance, *these eight identities should be memorized.* Indeed, the student should be so familiar with them that he will be able to recognize them even when they are in slightly different forms. In particular, an identity should be recognizable when it is solved for any function occurring in it. For example, the identity

$$\sin^2 \theta + \cos^2 \theta = 1$$

when solved for $\sin \theta$, becomes

$$\sin \theta = \pm \sqrt{1 - \cos^2 \theta}$$

The double sign is placed before the radical since $\sin \theta$ is positive for some angles and negative for others. We recall that $\sin \theta$ is positive for angles with the terminal side above the x axis and negative for angles with the terminal side below the x axis.

EXAMPLE 1   Given $\sin \theta = -\frac{1}{3}$ and θ in quadrant III, find the other functions of θ by use of the fundamental identities.

Solution   It may be observed that if the sine, cosine, and tangent of an angle are known, the other three functions can be had immediately from the reciprocal relations. To find $\cos \theta$, we use identity (13), and obtain

$$\cos^2 \theta = 1 - \sin^2 \theta = 1 - (-\tfrac{1}{3})^2 = 1 - \tfrac{1}{9} = \tfrac{8}{9}$$

and

$$\cos \theta = -\frac{2\sqrt{2}}{3}$$

We choose the negative square root because the cosine of a third-quadrant angle is negative. Next, from identity (11),

$$\tan \theta = \frac{\sin \theta}{\cos \theta} = \left(-\frac{1}{3}\right) \div \left(-\frac{2\sqrt{2}}{3}\right) = \frac{1}{2\sqrt{2}}$$

Taking the reciprocals of the sine, cosine, and tangent of θ, we write

$$\sin \theta = -\tfrac{1}{3} \qquad\qquad \csc \theta = -3$$

$$\cos \theta = -\frac{2\sqrt{2}}{3} \qquad \sec \theta = -\frac{3}{2\sqrt{2}}$$

$$\tan \theta = \frac{1}{2\sqrt{2}} \qquad \cot \theta = 2\sqrt{2}$$

EXAMPLE 2   Express each of the other trigonometric functions in terms of $\tan \theta$.

Solution   From identity (10), we have immediately

$$\cot \theta = \frac{1}{\tan \theta}$$

Solving $1 + \tan^2 \theta = \sec^2 \theta$ for $\sec \theta$, we obtain

$$\sec \theta = \pm\sqrt{1 + \tan^2 \theta} \qquad \text{and} \qquad \cos \theta = \pm\frac{1}{\sqrt{1 + \tan^2 \theta}}$$

Substituting $1/\tan \theta$ for $\cot \theta$ in $\csc^2 \theta = 1 + \cot^2 \theta$ gives

$$\csc^2 \theta = 1 + \frac{1}{\tan^2 \theta} = \frac{\tan^2 \theta + 1}{\tan^2 \theta}$$

Hence,

$$\csc \theta = \pm\frac{\sqrt{1 + \tan^2 \theta}}{\tan \theta} \qquad \text{and} \qquad \sin \theta = \pm\frac{\tan \theta}{\sqrt{1 + \tan^2 \theta}}$$

Exercise 4-1

1. Solve the equation $\cot\theta = \cos\theta/\sin\theta$ for (a) $\cos\theta$, (b) $\sin\theta$.
2. Solve the equation $\tan\theta = \sin\theta/\cos\theta$ for (a) $\sin\theta$, (b) $\cos\theta$.
3. Solve the equation $\sin^2\theta + \cos^2\theta = 1$ for (a) $\sin\theta$, (b) $\cos\theta$.
4. Solve the equation $1 + \tan^2\theta = \sec^2\theta$ for (a) $\tan\theta$, (b) $\sec\theta$.
5. Solve the equation $1 + \cot^2\theta = \csc^2\theta$ for (a) $\csc\theta$, (b) $\cot\theta$.

Tell in which quadrants θ may be in order that each of the following equations will be true.

6. $\sin\theta = -\sqrt{1 - \cos^2\theta}$
7. $\cos\theta = -\sqrt{1 - \sin^2\theta}$
8. $\cot\theta = \sqrt{\csc^2\theta - 1}$
9. $\sec\theta = \sqrt{1 + \tan^2\theta}$

By use of the fundamental identities, find the values of the other five functions in each problem 10 through 19.

10. $\sin\theta = \frac{4}{5}$, θ in Q I
11. $\cos\theta = +\frac{5}{13}$, θ in Q IV
12. $\cos\theta = -\frac{3}{4}$, θ in Q III
13. $\sin\theta = -\frac{2}{3}$, θ in Q III
14. $\tan\theta = \frac{5}{12}$, θ in Q I
15. $\cot\theta = -\frac{3}{4}$, θ in Q IV
16. $\sec\theta = \dfrac{\sqrt{5}}{\sqrt{2}}$, θ in Q IV
17. $\csc\theta = -\dfrac{3}{\sqrt{7}}$, θ in Q III
18. $\tan\theta = \dfrac{\sqrt{2}}{\sqrt{3}}$, θ in Q I
19. $\cot\theta = \frac{3}{2}$, θ in Q I

20. Express each of the other functions in terms of $\sin\theta$.
21. Express each of the other functions in terms of $\cos\theta$.
22. Express each of the other functions in terms of $\sec\theta$.

4-3   Equivalent trigonometric expressions

Frequently, it is desirable to change a trigonometric expression into an equivalent expression which is simpler, or into an equivalent expression which contains only a specified function. The original expression and the expression to which it is reduced must be identical; that is, the expressions are to be equal for all angles for which each expression is defined. The form of a trigonometric expression may be changed by substitutions from the fundamental identities. Also algebraic operations, such as combining fractions, making indicated multiplications and divisions, and factoring, will reduce an expression to another form. Usually, a combination of algebraic operations and substitutions from the fundamental identities is needed to simplify a given expression. We shall illustrate this with some examples.

EXAMPLE 1   Reduce the expression $\sin\theta\,(\csc\theta - \sin\theta)$ to $\cos^2\theta$.

Solution   We replace $\csc \theta$ by $1/\sin \theta$ and then perform the indicated multiplication. Thus, we have

$$\sin \theta \left(\csc \theta - \sin \theta \right) = \sin \theta \left(\frac{1}{\sin \theta} - \sin \theta \right)$$

$$= 1 - \sin^2 \theta$$

$$= \cos^2 \theta \qquad \text{by identity (13)}$$

EXAMPLE 2   Reduce the expression $\dfrac{1 + \cot^2 \theta}{1 + \tan^2 \theta}$ to $\cot^2 \theta$.

Solution

$$\frac{1 + \cot^2 \theta}{1 + \tan^2 \theta} = \frac{\csc^2 \theta}{\sec^2 \theta} \qquad \text{by identities (14) and (15)}$$

$$= \frac{1/\sin^2 \theta}{1/\cos^2 \theta} \qquad \text{by identities (8) and (9)}$$

$$= \frac{1}{\sin^2 \theta} \frac{\cos^2 \theta}{1} = \frac{\cos^2 \theta}{\sin^2 \theta}$$

$$= \cot^2 \theta \qquad \text{by identity (12)}$$

EXAMPLE 3   Reduce $\dfrac{\tan^2 \theta}{\sec \theta + 1}$ to $\sec \theta - 1$.

Solution   Since the result is in terms of the secant, we make a substitution for $\tan^2 \theta$. Thus we obtain

$$\frac{\tan^2 \theta}{\sec \theta + 1} = \frac{\sec^2 \theta - 1}{\sec \theta + 1} \qquad \text{by identity (14)}$$

$$= \frac{(\sec \theta + 1)(\sec \theta - 1)}{\sec \theta + 1} \qquad \text{by factoring}$$

$$= \sec \theta - 1$$

Exercise 4-2

Reduce the first expression to the second expression in each problem 1 through 22.

1. $\sin \theta \sec \theta$; $\tan \theta$
2. $\csc \theta \cos \theta$; $\cot \theta$
3. $\sin \theta \cot \theta$; $\cos \theta$
4. $\cos \theta \tan \theta$; $\sin \theta$
5. $\sec \theta \cot \theta \sin \theta$; 1
6. $\tan \theta \csc \theta \cos \theta$; 1
7. $\sec \theta \cot \theta$; $\csc \theta$
8. $\sec \theta \sin \theta$; $\tan \theta$
9. $\sin^2 \theta \cot^2 \theta = \cos^2 \theta$
10. $\sin^2 \theta \csc \theta$; $\sin \theta$

11. $\tan^2 \theta (1 + \cot^2 \theta)$; $\sec^2 \theta$

12. $\tan^2 \theta (1 - \csc^2 \theta)$; -1

13. $\cos^2 \theta (\sec^2 \theta - 1)$; $\sin^2 \theta$

14. $\sec^2 \theta (1 - \sin^2 \theta)$; 1

15. $\cot \theta (\tan \theta + \cot \theta)$; $\csc^2 \theta$;

16. $\cos \theta (\sec \theta - \cos \theta)$; $\sin^2 \theta$

17. $\dfrac{\cos^2 \theta}{1 - \sin \theta}$; $\dfrac{\csc \theta + 1}{\csc \theta}$

18. $\dfrac{\sin^2 \theta}{1 + \cos \theta}$; $1 - \cos \theta$

19. $\dfrac{\csc^2 \theta - 1}{\csc \theta + 1}$; $\csc \theta - 1$

20. $\dfrac{1 - \cos^2 \theta}{1 + \tan^2 \theta}$; $\cos^2 \theta \sin^2 \theta$

21. $\dfrac{\tan^2 \theta}{\sec \theta + 1}$; $\sec \theta - 1$

22. $\dfrac{1 - \sin^2 \theta}{1 + \cot^2 \theta}$; $\sin^2 \theta \cos^2 \theta$

4-4 Proving identities

In the preceding section we transformed trigonometric expressions into other equivalent expressions. The algebraic operations and the substitutions altered the form but not the value of the expression. This means in each case that the original expression and final expression, as well as the intermediate expressions, are identically equal. Any two of the expressions could be written as the members of an equation.

We shall next consider the problem of proving identities, that is, of proving that one member of an equation is identically equal to the other member. A trigonometric identity is really a theorem, and the process of establishing the theorem consists in making logical deductions from known facts, just as proofs are made in geometry. The tools which we have for this purpose are the fundamental identities and algebraic operations. If by these means we reduce one member of a trigonometric equation to the form of the other, then the identity (or theorem) is established.

It is not possible to give specific rules for proving identities. However, we make the following suggestions.

1. It is usually better to change the more complicated member of the equation into the form of the other member.
2. If one member contains only one function of an angle, it might be of advantage to express the functions in the other member in terms of the single function.
3. Usually it is preferable to avoid substitutions from the fundamental identities which introduce radicals.
4. Frequently it is helpful to combine fractions, perform indicated multiplications, or make divisions. In other cases, factoring is useful, and sometimes a fraction needs to be written as the sum of two or more fractions.
5. While often not the most direct way, many reductions can be made satisfactorily by first changing all the functions to sines and cosines.

6. Sometimes the proof can be facilitated by multiplying the numerator and denominator of a fraction by the same factor.
7. In all steps of the process, the member which is not being altered should be kept in view. This member is the goal, and reductions leading to it must be discovered.

EXAMPLE 1 Prove that the equation

$$(1 + \sin \theta)(1 - \sin \theta) = \cos^2 \theta$$

is an identity.

Solution Performing the multiplication indicated in the left member and using the fundamental identity (13), we have

$$(1 + \sin \theta)(1 - \sin \theta) = 1 - \sin^2 \theta = \cos^2 \theta$$

We have transformed one member of the given equation to the form of the other member; hence the equation is an identity.

EXAMPLE 2 Prove the identity

$$\frac{\cos^2 \theta + \tan^2 \theta - 1}{\sin^2 \theta} = \tan^2 \theta$$

Solution We shall work with the left member since it is more complicated than the right member. First, expressing the left member as the sum of two fractions, we have

$$\frac{\cos^2 \theta + \tan^2 \theta - 1}{\sin^2 \theta} = \frac{\cos^2 \theta - 1}{\sin^2 \theta} + \frac{\tan^2 \theta}{\sin^2 \theta}$$

$$= \frac{-\sin^2 \theta}{\sin^2 \theta} + \frac{\sin^2 \theta}{\cos^2 \theta} \frac{1}{\sin^2 \theta} \qquad \text{by identities (13) and (11)}$$

$$= -1 + \sec^2 \theta \qquad \text{by identity (9)}$$

$$= \tan^2 \theta \qquad \text{by identity (14)}$$

We have transformed one member of the equation to the form of the other; hence, the given equation is an identity.

EXAMPLE 3 Prove the identity

$$\frac{3 + \csc \theta}{3 - \csc \theta} = \frac{3 \sin \theta + 1}{3 \sin \theta - 1}$$

Solution We establish the identity immediately by multiplying the numerator and denominator of the left member of the given equation by $\sin \theta$. Thus

$$\frac{3 + \csc \theta}{3 - \csc \theta} = \frac{3 \sin \theta + \sin \theta \csc \theta}{3 \sin \theta - \sin \theta \csc \theta}$$

$$= \frac{3 \sin \theta + 1}{3 \sin \theta - 1} \qquad \text{by identity (8)}$$

EXAMPLE 4   Prove the identity

$$\frac{\sec \theta + 1}{\tan \theta} = \frac{\tan \theta}{\sec \theta - 1}$$

Solution   Let us express the right side of this equation as a fraction with the denominator like that of the left side. As a first step toward this end, we multiply the numerator and denominator of the right side by $\tan \theta$. Thus we get

$$\frac{\tan \theta}{\sec \theta - 1} = \frac{\tan^2 \theta}{(\sec \theta - 1) \tan \theta}$$

$$= \frac{\sec^2 \theta - 1}{(\sec \theta - 1) \tan \theta} \qquad \text{by identity (14)}$$

$$= \frac{(\sec \theta - 1)(\sec \theta + 1)}{(\sec \theta - 1) \tan \theta} \qquad \text{by factoring}$$

$$= \frac{\sec \theta + 1}{\tan \theta}$$

EXAMPLE 5   Prove the identity

$$\frac{1}{1 - \cos \theta} - \frac{1}{1 + \cos \theta} = 2 \cot \theta \csc \theta$$

Solution   Combining the fractions of the left member, we get

$$\frac{1}{1 - \cos \theta} - \frac{1}{1 + \cos \theta} = \frac{(1 + \cos \theta) - (1 - \cos \theta)}{(1 - \cos \theta)(1 + \cos \theta)}$$

$$= \frac{2 \cos \theta}{1 - \cos^2 \theta}$$

$$= \frac{2 \cos \theta}{\sin^2 \theta} \qquad \text{by identity (13)}$$

$$= \frac{2 \cos \theta}{\sin \theta} \frac{1}{\sin \theta}$$

$$= 2 \cot \theta \csc \theta \qquad \text{by identities (8) and (12)}$$

EXAMPLE 6   Prove the identity

$$\frac{\tan \theta - \cot \theta}{\tan \theta + \cot \theta} = 1 - 2 \cos^2 \theta$$

Solution We proceed by expressing the left member of the equation in terms of sines and cosines. Thus

$$\frac{\tan\theta - \cot\theta}{\tan\theta + \cot\theta} = \frac{\dfrac{\sin\theta}{\cos\theta} - \dfrac{\cos\theta}{\sin\theta}}{\dfrac{\sin\theta}{\cos\theta} + \dfrac{\cos\theta}{\sin\theta}} \qquad \text{by identities (11) and (12)}$$

$$= \frac{\sin^2\theta - \cos^2\theta}{\sin^2\theta + \cos^2\theta} \qquad \text{multiplying numerator and denominator by } \sin\theta\cos\theta$$

$$= \frac{\sin^2\theta - \cos^2\theta}{1} \qquad \text{by identity (13)}$$

$$= 1 - \cos^2\theta - \cos^2\theta \qquad \text{by identity (13)}$$

$$= 1 - 2\cos^2\theta$$

Exercise 4-3

Prove that the following equations are identities.

1. $(1 + \cos\theta)(1 - \cos\theta) = \sin^2\theta$
2. $(\sec\theta + 1)(\sec\theta - 1) = \tan^2\theta$
3. $\csc^2\theta(1 - \sin^2\theta) = \cot^2\theta$
4. $\cot\theta(\cot\theta - \tan\theta) = \csc^2\theta - 2$
5. $(\sin^2\theta - \cos^2\theta) = 2\sin^2\theta - 1$
6. $\cos\theta(\sec\theta - \cos\theta) = \sin^2\theta$
7. $(\sin\theta - \cos\theta)^2 = 1 - 2\sin\theta\cos\theta$
8. $\sin^2\theta - \cos^2\theta = 1 - 2\cos^2\theta$
9. $(\sec\theta + 1)(\cos\theta - 1) = \cos\theta - \sec\theta$
10. $(\cot\theta + \tan\theta)^2 = \sec^2\theta + \csc^2\theta$

11. $\dfrac{\tan\theta - \cos\theta}{\sin\theta} = \sec\theta - \cot\theta$

12. $\dfrac{\sin\theta - \cos\theta}{\sin\theta} = 1 - \cot\theta$

13. $\dfrac{\sin^2\theta - \cos^2\theta}{\sin\theta\cos\theta} = \tan\theta - \cot\theta$

14. $\dfrac{\sin\theta - \sec\theta}{\cos\theta} = \tan\theta - \sec^2\theta$

15. $\dfrac{\sin^2\theta + \tan^2\theta}{\sin^2\theta} = 1 + \sec^2\theta$

16. $\dfrac{\cos^2\theta + \cot^2\theta}{\cos^2\theta} = 1 + \csc^2\theta$

17. $\dfrac{\cos^2\theta + \sin^2\theta}{\sin\theta\,\cos\theta} = \sec\theta\,\csc\theta$

18. $\dfrac{\tan\theta - \cot\theta}{\sin\theta\,\cos\theta} = \sec^2\theta - \csc^2\theta$

19. $\sin^2\theta - \sec^2\theta = -\cos^2\theta - \tan^2\theta$

20. $\cot^2\theta = \dfrac{\sin^2\theta + \cot^2\theta - 1}{\cos^2\theta}$

21. $\dfrac{1 + 2\tan\theta}{1 - 2\tan\theta} = \dfrac{\cot\theta + 2}{\cot\theta - 2}$

22. $\dfrac{\cos\theta - 1}{\cos\theta + 1} = \dfrac{1 - \sec\theta}{1 + \sec\theta}$

23. $\dfrac{\cos\theta + \sin\theta}{\cos\theta - \sin\theta} = \dfrac{\csc\theta + \sec\theta}{\csc\theta - \sec\theta}$

24. $\dfrac{\cot\theta + 1}{\cot\theta - 1} = \dfrac{\cos\theta + \sin\theta}{\cos\theta - \sin\theta}$

25. $\dfrac{1 + \cos^2\theta}{\sin^2\theta} = \dfrac{\sec\theta + \cos\theta}{\sec\theta - \cos\theta}$

26. $\dfrac{\cot\theta + 2}{\cot\theta - 3} = \dfrac{\cos\theta + 2\sin\theta}{\cos\theta - 3\sin\theta}$

27. $\dfrac{\sin^2\theta}{\sin\theta - 1} = \dfrac{1 + \cos\theta}{\sec\theta}$

28. $\dfrac{\sin\theta + \cos\theta}{\sec\theta - \cos\theta} = \dfrac{1 + \cot\theta}{\csc\theta\,\sec\theta - \cot\theta}$

29. $\dfrac{\cos\theta}{1 - \sin\theta} = \dfrac{1 + \sin\theta}{\cos\theta}$

30. $\dfrac{\sec\theta - 1}{\cos\theta} = \dfrac{\tan^2\theta}{\cos\theta + 1}$

31. $\dfrac{2 - \sec^2\theta}{\cot\theta - 1} = \dfrac{1 + \tan\theta}{\cot\theta}$

32. $\dfrac{1 + \cos\theta}{\sin\theta} = \dfrac{\sin\theta}{1 - \cos\theta}$

33. $\dfrac{1}{\tan\theta - \sec\theta} - \dfrac{1}{\tan\theta + \sec\theta} = -2\sec\theta$

34. $\dfrac{1}{1 - \sin\theta} - \dfrac{1}{1 + \sin\theta} = 2\sin\theta\,\sec^2\theta$

35. $\dfrac{1 + \cos\theta}{1 - \cos\theta} - \dfrac{1 - \cos\theta}{1 + \cos\theta} = 4\cos\theta\,\csc^2\theta$

36. $\dfrac{1}{\csc\theta - \cos\theta} - \dfrac{1}{\csc\theta + \cos\theta} = 2\cot\theta$

37. $\dfrac{\cos^2 \theta}{\cot^2 \theta - \cos^2 \theta} = \tan^2 \theta$

38. $\dfrac{\sin^2 \theta}{\tan^2 \theta - \sin^2 \theta} = \cot^2 \theta$

39. $\cot \theta + \tan \theta = \csc \theta \sec \theta$
40. $\tan \theta \sin \theta + \cos \theta = \sec \theta$
41. $\csc^2 \theta + \sec^2 \theta = \csc^2 \theta \sec^2 \theta$
42. $\tan^2 \theta - \sin^2 \theta = \tan^2 \theta \sin^2 \theta$
43. $\cot^2 \theta - \cos^2 \theta = \cot^2 \theta \cos^2 \theta$
44. $\csc \theta - \cot \theta \cos \theta = \sin \theta$
45. $\sec \theta - \tan \theta \sin \theta = \cos \theta$
46. $\csc \theta - \sin \theta = \cos \theta \cot \theta$
47. $\cot \theta + \tan \theta = \csc \theta \sec \theta$
48. $(\csc \theta - \cot \theta)(\cos \theta + 1) = \sin \theta$
49. $(\sec \theta + \csc \theta)(\sin \theta - \cos \theta) = \tan \theta - \cot \theta$
50. $(\sec \theta + \tan \theta)(\cot \theta - \cos \theta) = \csc \theta - \sin \theta$
51. $(\csc \theta + \cot \theta)(\tan \theta - \sin \theta) = \sec \theta - \cos \theta$
52. $\cot^2 \theta + \sin^2 \theta + \cos^2 \theta = \csc^2 \theta$
53. $\cos^2 \theta + \tan^2 \theta + \sin^2 \theta = \sec^2 \theta$

54. $\dfrac{1 + \cot^2 \theta}{1 + \tan^2 \theta} = \cot^2 \theta$

55. $\dfrac{\cos \theta}{\sec \theta} + \dfrac{\sin \theta}{\csc \theta} = \sin \theta \csc \theta$

56. $\tan^2 \theta + \csc^2 \theta = \cot^2 \theta + \sec^2 \theta$
57. $\sec \theta + \csc \theta = (\sin \theta + \cos \theta)(\tan \theta + \cot \theta)$

58. $\cos^2 \theta - \csc^2 \theta = \dfrac{\sin^2 \theta - \sec^2 \theta}{\tan^2 \theta}$

59. $\dfrac{1 + \cot^2 \theta}{1 - \sin^2 \theta} = \csc^2 \theta \sec^2 \theta$

60. $\dfrac{\sec^2 \theta - 1}{\sec^2 \theta} = \sin^2 \theta$

61. $\dfrac{\cot^2 \theta + \sec^2 \theta + 1}{\cot^2 \theta} = \sec^4 \theta$

62. $\dfrac{1 + \tan^2 \theta + \csc^2 \theta}{\sec^2 \theta} = \csc^2 \theta$

63. $\dfrac{\cos^2 \theta - \sin^2 \theta}{\cos \theta + \sin \theta} = \cos \theta - \sin \theta$

64. $\dfrac{\cos^4 \theta - \sin^4 \theta}{\cos^2 \theta - \sin^2 \theta} = 1$

65. $\dfrac{\tan^3 \theta - 1}{\tan \theta - 1} = \tan \theta + \sec^2 \theta$

66. $\dfrac{\cot^3 \theta + 1}{\cot \theta + 1} = \csc^2 \theta - \cot \theta$

4-5 Cosine of the difference and sum of two angles

The eight fundamental identities (Sec. 4-2) involve functions of a single angle. In the remainder of this chapter, we shall consider identities involving functions of more than one angle. We start by proving that the cosine of the difference and sum of two angles are given by the formulas

$$\cos(A - B) = \cos A \cos B + \sin A \sin B \qquad (16)$$
$$\cos(A + B) = \cos A \cos B - \sin A \sin B \qquad (17)$$

The angles A and B (Fig. 4-2) are constructed in standard position, so that angle P_1OP_2 is equal to $B - A$. We construct angle P_0OP_3 the same size as $B - A$ but measured clockwise. Hence, angle $P_0OP_3 = A - B$. It is then evident that chords P_1P_2 and P_0P_3 have the same length. Recalling that the terminal side of an angle cuts the unit circle at a point whose coordinates are the cosine and sine of the angle (Fig. 4-1), we see that the terminal sides of A, B, and $A - B$ cut the unit circle at the points whose coordinates are, respectively,

$P_1(\cos A,\ \sin A)$
$P_2(\cos B,\ \sin B)$
$P_3[\cos(A - B),\ \sin(A - B)]$

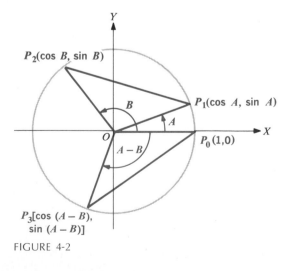

FIGURE 4-2

Since the chords P_1P_2 and P_0P_3 have the same length, we may apply the distance formula (Sec. 1-5), and write

$$(P_1P_2)^2 = (\cos B - \cos A)^2 + (\sin B - \sin A)^2$$
$$= 2 - 2(\cos A \cos B + \sin A \sin B)$$
$$(P_0P_3)^2 = [\cos(A - B) - 1]^2 + \sin^2(A - B)$$
$$= \cos^2(A - B) - 2\cos(A - B) + 1 + \sin^2(A - B)$$
$$= 2 - 2\cos(A - B)$$

We now equate the quantities which are equal to the squares of the length of P_1P_2 and P_0P_3 and have

$$\cos(A - B) = \cos A \cos B + \sin A \sin B \tag{16}$$

The method of deriving this formula will apply regardless of the measures of A and B. Hence, the formula holds where A and B are any angles whatever, positive, zero, or negative. We state the formula in words:

The cosine of the difference of two angles is equal to the cosine of the first times the cosine of the second plus the sine of the first times the sine of the second.

Before establishing the formula for the cosine of the sum of two angles, we use formula (16) to develop some additional identities. Replacing A by $\pi/2$ and recalling that $\cos \pi/2 = 0$ and $\sin \pi/2 = 1$, we have

$$\cos\left(\frac{\pi}{2} - B\right) = \cos\frac{\pi}{2}\cos B + \sin\frac{\pi}{2}\sin B$$

and therefore

$$\cos\left(\frac{\pi}{2} - B\right) = \sin B \tag{18}$$

Formula (18) holds for all values of B, and in particular when B is replaced by $\pi/2 - B$. Thus

$$\cos\left[\frac{\pi}{2} - \left(\frac{\pi}{2} - B\right)\right] = \sin\left(\frac{\pi}{2} - B\right)$$

Since $\pi/2 - (\pi/2 - B) = B$, we conclude that

$$\sin\left(\frac{\pi}{2} - B\right) = \cos B \tag{19}$$

If we let $A = 0$ in formula (16) and use $\cos 0 = 1$ and $\sin 0 = 0$, we get

$$\cos(0 - B) = \cos 0 \cos B + \sin 0 \sin B$$
$$\cos(-B) = \cos B \tag{20}$$

In formula (18), we replace B by $-B$ and find that

$$\sin (-B) = \cos \left(\frac{\pi}{2} + B \right)$$

$$= \cos \left[B - \left(-\frac{\pi}{2} \right) \right]$$

$$= \cos B \cos \left(-\frac{\pi}{2} \right) + \sin B \sin \left(-\frac{\pi}{2} \right)$$

Hence,

$$\sin (-B) = -\sin B \tag{21}$$

Formulas (20) and (21) now yield

$$\tan (-B) = -\tan B \tag{22}$$

We may now use formulas (20), (21), and (16) in finding the cosine of the sum of two angles. Thus,

$$\cos (A + B) = \cos [A - (-B)]$$
$$= \cos A \cos (-B) + \sin A \sin (-B)$$

Hence,

$$\cos (A + B) = \cos A \cos B - \sin A \sin B \tag{17}$$

EXAMPLE 1   Find the exact value of $\cos (\pi/12)$.

Solution   Observing that $\pi/12 = \pi/3 - \pi/4$, we apply formula (16) and get

$$\cos \frac{\pi}{12} = \cos \left(\frac{\pi}{3} - \frac{\pi}{4} \right)$$

$$= \cos \frac{\pi}{3} \cos \frac{\pi}{4} + \sin \frac{\pi}{3} \sin \frac{\pi}{4}$$

$$= \frac{1}{2} \frac{\sqrt{2}}{2} + \frac{\sqrt{3}}{2} \frac{\sqrt{2}}{2} = \frac{\sqrt{2} + \sqrt{6}}{4}$$

EXAMPLE 2   If $\sin A = \frac{3}{5}$ and $\cos B = \frac{5}{13}$ and if A is in quadrant II and B is in quadrant IV, find $\cos (A - B)$ and $\cos (A + B)$.

Solution   From the quadrants of A and B, $\cos A$ is negative and $\sin B$ is negative. Hence,

$$\cos A = -\sqrt{1 - \sin^2 A} = -\sqrt{1 - \tfrac{9}{25}} = -\tfrac{4}{5}$$
$$\sin B = -\sqrt{1 - \cos^2 B} = -\sqrt{1 - \tfrac{25}{169}} = -\tfrac{12}{13}$$

Applying formulas (16) and (17), we obtain

$$\cos (A - B) = \cos A \cos B + \sin A \sin B$$
$$= (-\tfrac{4}{5})(\tfrac{5}{13}) - (\tfrac{3}{5})(\tfrac{12}{13}) = -\tfrac{56}{65}$$

and

$$\cos (A + B) = \cos A \cos B - \sin A \sin B$$
$$= (-\tfrac{4}{5})(\tfrac{5}{13}) + (\tfrac{3}{5})(\tfrac{12}{13}) = \tfrac{16}{65}$$

In Example 2 we found the values of $\cos (A - B)$ and $\cos (A + B)$ from the known values of the cosines and sines of A and B separately. It is also important to recognize an expression in the form of the right side of formula (16) as the cosine of the difference of two angles and to recognize an expression in the form of the right member of formula (17) as the cosine of the sum of two angles. For example, the expression

$$\cos 3\theta \cos 2\theta - \sin 3\theta \sin 2\theta$$

is equal to $\cos (3\theta + 2\theta) = \cos 5\theta$, and the expression

$$\cos 70° \cos 25° + \sin 70° \sin 25°$$

is equal to $\cos (70° - 25°) = \cos 45°$.

Exercise 4-4

Use formulas (16) and (17) and find, without tables, the value of the cosine of each of the following angles.

1. 75°   **2.** $-15°$ **3.** $-75°$ **4.** 105°
5. 165°   **6.** 195°   **7.** 285°   **8.** 255°

9. Show that $\cos (45° + 30°)$ is not equal to $\cos 45° + \cos 30°$.
10. Show that $\cos (60° + 45°)$ is not equal to $\cos 60° + \cos 45°$.
11. Show that $\cos 180°$ is not equal to $2 \cos 90°$.
12. Show that $\cos (180° - 30°)$ is not equal to $\cos 180° - \cos 30°$.

By means of formulas (16) and (17) express each of the following quantities as a function of θ.

13. $\cos (\pi + \theta)$ **14.** $\cos (90° + \theta)$ **15.** $\cos (\pi - \theta)$
16. $\cos (2\pi + \theta)$ **17.** $\cos (270° - \theta)$ **18.** $\cos (3\pi - \theta)$

Find the value of each expression.

19. $\cos 140° \cos 40° - \sin 140° \sin 40°$
20. $\cos 84° \cos 24° + \sin 84° \sin 24°$
21. $\cos 160° \cos 70° + \sin 160° \sin 70°$
22. $\cos (50° + \theta) \cos (20° + \theta) + \sin (50° + \theta) \sin (20° + \theta)$

Reduce each expression to a single term.

23. $\cos 3A \cos 2A - \sin 3A \sin 2A$

24. $\cos 4A \cos A + \sin 4A \sin A$
25. $\cos(60° - A) - \cos(60° + A)$
26. $\cos(45° - A) + \cos(45° + A)$

In problems 27 through 30 find the values of $\cos(A - B)$ and $\cos(A + B)$

27. A and B in Q I, $\cos A = \frac{3}{5}$, $\sin B = \frac{12}{13}$
28. A and B in Q II, $\sin A = +\frac{7}{25}$, $\cos B = -\frac{3}{5}$
29. A in Q III, B in Q IV, $\tan A = \frac{15}{8}$, $\tan B = -\frac{3}{4}$
30. A and B in Q II, $\sin A = \frac{8}{17}$, $\sin B = \frac{3}{5}$

31. Use Fig. 4-3 and show that $OS = \cos(A + B)$, $OT = \cos A \cos B$, $ST = \sin A \sin B$ and consequently $\cos(A + B) = \cos A \cos B - \sin A \sin B$.

32. Theorem   *Changing the sign of an angle does not change the value of its cosine and secant. Changing the sign of an angle reverses the sign of the sine and cosecant, and the tangent and cotangent.*

Use identities (20) to (22) and prove this theorem. Establish the theorem also by referring to Fig. 4-4. Notice that the theorem is true for any value of θ and $-\theta$ except for the undefined functions of a quadrantal angle.

4-6 Sine and tangent of the sum and difference of two angles

We now apply formulas (18) and (19) of the previous section to derive the formulas

$$\sin(A + B) = \sin A \cos B + \cos A \sin B \tag{23}$$
$$\sin(A - B) = \sin A \cos B - \cos A \sin B \tag{24}$$

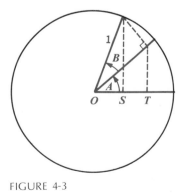

FIGURE 4-3

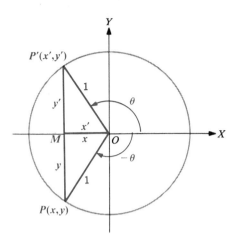

FIGURE 4-4

Replacing B by $(A + B)$ in (18) and reversing the members of the equation gives

$$\sin (A + B) = \cos \left[\frac{\pi}{2} - (A + B)\right] = \cos \left[\left(\frac{\pi}{2} - A\right) - B\right]$$

$$= \cos \left(\frac{\pi}{2} - A\right) \cos B + \sin \left(\frac{\pi}{2} - A\right) \sin B$$

Then, using (18) and (19) to change the form of the right member of this equation, we have

$$\sin (A + B) = \sin A \cos B + \cos A \sin B \qquad (23)$$

To obtain the sine of the difference of two angles, we replace B by $-B$ in (23). This gives

$$\sin [A + (-B)] = \sin A \cos (-B) + \cos A \sin (-B)$$

Formulas (20) and (21) permit us to write

$$\sin (A - B) = \sin A \cos B - \cos A \sin B \qquad (24)$$

Formulas (16), (17), (23), and (24) enable us to obtain the formulas

$$\tan (A + B) = \frac{\tan A + \tan B}{1 - \tan A \tan B} \qquad (25)$$

$$\tan (A - B) = \frac{\tan A - \tan B}{1 + \tan A \tan B} \qquad (26)$$

Since $\tan \theta = \sin \theta / \cos \theta$, we have

$$\tan (A + B) = \frac{\sin (A + B)}{\cos (A + B)} = \frac{\sin A \cos B + \cos A \sin B}{\cos A \cos B - \sin A \sin B}$$

If we divide each term of the numerator and each term of the denominator of the right member of this equation by cos A cos B and simplify, we obtain formula (25). Formula (26) can be derived in a similar way.

EXAMPLE 1   Find tan 15° by using tangents of 60° and 45°.

Solution   Since $15° = 60° - 45°$, by formula (26), we have

$$\tan 15° = \tan (60° - 45°)$$
$$= \frac{\tan 60° - \tan 45°}{1 + \tan 60° \tan 45°}$$
$$= \frac{\sqrt{3} - 1}{1 + \sqrt{3}} = 2 - \sqrt{3}$$

EXAMPLE 2   If A is in quadrant III, B is in quadrant IV, $\tan A = \frac{8}{15}$, and $\tan B = -\frac{5}{12}$, find the sine and tangent of $(A + B)$ and $(A - B)$.

Solution   The two angles are constructed in Fig. 4-5. Reading the values of the sine, cosine, and tangent of angles A and B from the figure, we find

$$\sin (A + B) = \left(-\frac{8}{17}\right)\left(\frac{12}{13}\right) + \left(-\frac{15}{17}\right)\left(-\frac{5}{13}\right)$$
$$= \frac{-96 + 75}{(17)(13)} = -\frac{21}{221}$$
$$\sin (A - B) = \frac{-96 - 75}{221} = -\frac{171}{221}$$
$$\tan (A + B) = \frac{\frac{8}{15} + (-\frac{5}{12})}{1 - (\frac{8}{15})(-\frac{5}{12})} = \frac{96 - 75}{180 + 40} = \frac{21}{220}$$
$$\tan (A - B) = \frac{96 + 75}{180 - 40} = \frac{171}{140}$$

From the signs of these results we see that the terminal side of $(A + B)$ is in quadrant III and the terminal side of $(A - B)$ is also in quadrant III.

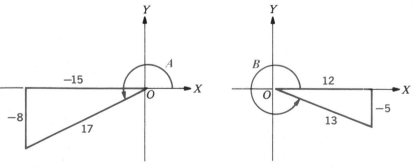

FIGURE 4-5

Exercise 4-5

By means of formulas (23) through (26) find, without tables, the values of the sine and tangent of each angle.

1. $75°$　　　**2.** $-15°$　　　**3.** $-75°$　　　**4.** $105°$
5. $225°$　　　**6.** $-195°$　　　**7.** $165°$　　　**8.** $-285°$

9. Show that $\sin(45° - 30°)$ is not equal to $\sin 45° - \sin 30°$.
10. Show that $\sin(60° + 45°)$ is not equal to $\sin 60° + \sin 45°$.
11. Show that $\sin(180° - 30°)$ is not equal to $\sin 180° - \sin 30°$.
12. Show that $\tan 60°$ is not equal to $2 \tan 30°$.
13. Show that the reduction formulas of Sec. 3-4 involving θ are valid for θ any angle whatever.

Find the value of each expression.

14. $\sin 70° \cos 20° + \cos 70° \sin 20°$
15. $\sin 50° \cos 20° - \cos 50° \sin 20°$
16. $\sin 160° \cos 115° - \cos 160° \sin 115°$

17. $\dfrac{\tan 44° + \tan 16°}{1 - \tan 44° \tan 16°}$　　　**18.** $\dfrac{\tan 70° - \tan 40°}{1 + \tan 70° \tan 40°}$

Reduce each expression to a single term.

19. $\sin 4\theta \cos \theta + \cos 4\theta \sin \theta$
20. $\sin(30° + \theta) - \sin(30° - \theta)$

21. $\dfrac{\tan 5\theta + \tan \theta}{1 - \tan 5\theta \tan \theta}$　　　**22.** $\dfrac{\tan 6A + \tan 5A}{1 - \tan 6A \tan 5A}$

By means of formulas (23) and (24) express each of the following quantities as a function of θ.

23. $\sin(90° + \theta)$　　**24.** $\sin(\pi + \theta)$　　**25.** $\sin(\pi - \theta)$
26. $\sin(270° - \theta)$　　**27.** $\sin(3\pi - \theta)$　　**28.** $\sin(2\pi + \theta)$

Verify the following identities.

29. $\dfrac{\sin(x - y)}{\sin(x + y)} = \dfrac{\tan x - \tan y}{\tan x + \tan y}$

30. $\dfrac{\cos(x - y)}{\sin(x + y)} = \dfrac{1 + \tan x \tan y}{\tan x + \tan y}$

31. $\dfrac{\cos(x + y)}{\cos(x - y)} = \dfrac{1 - \tan x \tan y}{1 + \tan x \tan y}$

32. $\dfrac{\tan(45° + x)}{\tan(45° - x)} = \dfrac{(1 + \tan x)^2}{(1 - \tan x)^2}$

33. $\sin (A + B + C) = \sin A \cos B \cos C + \cos A \sin B \cos C$
$\qquad + \cos A \cos B \sin C - \sin A \sin B \sin C$

34. $\cos (A + B + C) = \cos A \cos B \cos C - \sin A \sin B \cos C$
$\qquad - \sin A \cos B \sin C - \cos A \sin B \sin C$

In problems 35 through 38 find the values of the sine and tangent of $(A + B)$ and of $(A - B)$. Give the quadrant in which $(A + B)$ terminates and the quadrant in which $(A - B)$ terminates.

35. A and B in Q I, $\sin A = \frac{5}{13}$, $\sin B = \frac{3}{5}$
36. A in Q II, B in Q I, $\sin A = \frac{7}{25}$, $\sin B = \frac{4}{5}$
37. A in Q IV, B in Q III, $\tan A = -\frac{15}{8}$, $\tan B = +\frac{3}{4}$
38. A in Q III, B in Q II, $\tan B = -2$, $\sin A = -\frac{4}{5}$

39. Using formula (25) and the identity $\cot \theta = 1/\tan \theta$, show that

$$\cot (A + B) = \frac{\cot A \cot B - 1}{\cot B + \cot A}$$

40. Use $\sin (A - B)$ and $\cos (A - B)$ and derive the formula

$$\cot (A - B) = \frac{\cot A \cot B + 1}{\cot B - \cot A}$$

41. Locate line segments in Fig. 4-3 which are equal to $\sin (A + B)$, $\sin A \cos B$, and $\cos A \sin B$. Using these line segments, show that $\sin (A + B) = \sin A \cos B + \cos A \sin B$.

4-7 Functions of twice an angle

Formulas for the functions of twice an angle may be had readily from the functions of the sum of two angles. Setting $B = A$ in formulas (23), (17), and (25), we obtain the following results.

$$\sin (A + A) = \sin A \cos A + \cos A \sin A$$
$$\cos (A + A) = \cos A \cos A - \sin A \sin A$$
$$\tan (A + A) = \frac{\tan A + \tan A}{1 - \tan A \tan A}$$

Hence

$$\sin 2A = 2 \sin A \cos A \qquad (27)$$
$$\cos 2A = \cos^2 A - \sin^2 A$$
$$= 1 - 2 \sin^2 A$$
$$= 2 \cos^2 A - 1 \qquad (28)$$

$$\tan 2A = \frac{2 \tan A}{1 - \tan^2 A} \qquad (29)$$

Formulas (27) to (29) are sometimes called the *double-angle formulas*. The student should state these formulas in words. It is essential to notice that formulas (27) and (28) are true for all angles A, and formula (29) is true for all angles except those for which $\tan A = \pm 1$. With this exception each of the formulas is valid if the angle in the left member is twice the angle in the right member. Thus, as illustrations, we write

$$\sin 20° = 2 \sin 10° \cos 10°$$

$$\cos 4x = \cos^2 2x - \sin^2 2x$$

$$\tan \theta = \frac{2 \tan \frac{1}{2}\theta}{1 - \tan^2 \frac{1}{2}\theta}$$

EXAMPLE 1 Find the functions of $2x$ if x is in quadrant II and $\sin x = \frac{3}{5}$.

Solution First we notice that $\cos x = -\frac{4}{5}$ and $\tan x = -\frac{3}{4}$. Then from formulas (27) to (29), we get

$$\sin 2x = 2 \sin x \cos x = 2(\tfrac{3}{5})(-\tfrac{4}{5}) = -\tfrac{24}{25}$$

$$\cos 2x = \cos^2 x - \sin^2 x = (-\tfrac{4}{5})^2 - (\tfrac{3}{5})^2 = \tfrac{7}{25}$$

$$\tan 2x = \frac{2 \tan x}{1 - \tan^2 x} = \frac{2(-\tfrac{3}{4})}{1 - (-\tfrac{3}{4})^2} = -\frac{24}{7}$$

The other three functions may be obtained as the reciprocals of the three here computed.

EXAMPLE 2 Express $\sin 3\theta$ in terms of $\sin \theta$.

Solution First, we write $3\theta = 2\theta + \theta$; then from formulas (23), (27), and (28) we obtain

$$\begin{aligned}
\sin (2\theta + \theta) &= \sin 2\theta \cos \theta + \cos 2\theta \sin \theta \\
&= (2 \sin \theta \cos \theta) \cos \theta + (1 - 2 \sin^2 \theta) \sin \theta \\
&= 2 \sin \theta \cos^2 \theta + \sin \theta - 2 \sin^3 \theta \\
&= 2 \sin \theta (1 - \sin^2 \theta) + \sin \theta - 2 \sin^3 \theta \\
&= 3 \sin \theta - 4 \sin^3 \theta
\end{aligned}$$

Hence,

$$\sin 3\theta = 3 \sin \theta - 4 \sin^3 \theta$$

EXAMPLE 3 Prove the identity $\dfrac{\sin 2x}{1 - \cos 2x} = \cot x$.

Solution   From (27) and the second form of (28), we have

$$\frac{\sin 2x}{1 - \cos 2x} = \frac{2 \sin x \cos x}{1 - (1 - 2 \sin^2 x)}$$

$$= \frac{2 \sin x \cos x}{2 \sin^2 x} = \frac{\cos x}{\sin x} = \cot x$$

4-8 Functions of half an angle

Frequently it is desirable to obtain the functions of half an angle from the known functions of the whole angle. For this purpose the second and third forms of formula (28) may be conveniently used. To emphasize the fact that the angle in the right member is half the angle appearing in the left member, we replace A by $\frac{1}{2}A$ and $2A$ by A and write the identities

$$\cos A = 1 - 2 \sin^2 \tfrac{1}{2}A \qquad \text{and} \qquad \cos A = 2 \cos^2 \tfrac{1}{2}A - 1$$

We solve the first of the equations for $\sin \frac{1}{2}A$ and the second for $\cos \frac{1}{2}A$. Thus we get

$$2 \sin^2 \tfrac{1}{2}A = 1 - \cos A$$

$$\sin^2 \tfrac{1}{2}A = \frac{1 - \cos A}{2}$$

$$\sin \tfrac{1}{2}A = \pm \sqrt{\frac{1 - \cos A}{2}} \tag{30}$$

And similarly,

$$\cos \tfrac{1}{2}A = \pm \sqrt{\frac{1 + \cos A}{2}} \tag{31}$$

We next divide the members of identity (30) by the corresponding members of identity (31) and obtain

$$\tan \tfrac{1}{2}A = \pm \sqrt{\frac{1 - \cos A}{1 + \cos A}} \tag{32}$$

The double signs are placed before the radicals of formulas (30) to (32) because $\frac{1}{2}A$ may terminate in any of the four quadrants. The proper sign for each function must be chosen when the quadrant of $\frac{1}{2}A$ is determined. Thus if $\frac{1}{2}A$ is in the first or second quadrant, $\sin \frac{1}{2}A$ is positive; but if $\frac{1}{2}A$ is a third- or fourth-quadrant angle, then $\sin \frac{1}{2}A$ is negative. For example, if $180° \leq A \leq 360°$ so that $90° \leq \frac{1}{2}A \leq 180°$, then

$$\sin \tfrac{1}{2}A = \sqrt{\frac{1 - \cos A}{2}} \qquad \text{and} \qquad \cos \tfrac{1}{2}A = -\sqrt{\frac{1 + \cos A}{2}}$$

Two rational expressions for $\tan \frac{1}{2}A$ may be derived as follows:

$$\tan \tfrac{1}{2}A = \frac{\sin \tfrac{1}{2}A}{\cos \tfrac{1}{2}A} = \frac{\sin \tfrac{1}{2}A}{\cos \tfrac{1}{2}A}\frac{2 \sin \tfrac{1}{2}A}{2 \sin \tfrac{1}{2}A} = \frac{2 \sin^2 \tfrac{1}{2}A}{2 \sin \tfrac{1}{2}A \cos \tfrac{1}{2}A}$$

Applying formulas (27) and (30), we have

$$\tan \tfrac{1}{2}A = \frac{1 - \cos A}{\sin A} \tag{33}$$

Similarly,

$$\tan \tfrac{1}{2}A = \frac{\sin A}{1 + \cos A} \tag{34}$$

EXAMPLE 1   Find $\sin 105°$ and $\tan 105°$.

Solution   If we use formula (30) with $\frac{1}{2}A = 105°$ and $A = 210°$, we have

$$\sin 105° = +\sqrt{\frac{1 - \cos 210°}{2}}$$

The positive sign is chosen because the sine of a second-quadrant angle is positive. Since $\cos 210° = -\frac{1}{2}\sqrt{3}$, we obtain

$$\sin 105° = \sqrt{\frac{1 + \tfrac{1}{2}\sqrt{3}}{2}} = \sqrt{\frac{2 + \sqrt{3}}{4}} = \frac{\sqrt{2 + \sqrt{3}}}{2}$$

To calculate $\tan 105°$, we may use formula (32), (33), or (34). Substituting in (33), we get

$$\tan 105° = \frac{1 - \cos 210°}{\sin 210°} = \frac{1 + \tfrac{1}{2}\sqrt{3}}{-\tfrac{1}{2}} = -(2 + \sqrt{3})$$

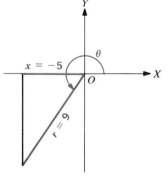

FIGURE 4-6

EXAMPLE 2   If θ is an angle in the third quadrant and $\cos \theta = -\frac{5}{9}$ (Fig. 4-6), find the values of the sine, cosine, and tangent of $\frac{1}{2}\theta$.

Solution   Since $\frac{1}{2}\theta$ is between $90°$ and $135°$, $\sin \frac{1}{2}\theta$ is positive and the cosine and tangent of $\frac{1}{2}\theta$ are negative. Then substituting in formulas (30) to (32), we obtain

$$\sin \tfrac{1}{2}\theta = \sqrt{\frac{1 + \frac{5}{9}}{2}} = \sqrt{\frac{14}{18}} = \frac{\sqrt{7}}{3}$$

$$\cos \tfrac{1}{2}\theta = -\sqrt{\frac{1 - \frac{5}{9}}{2}} = -\sqrt{\frac{4}{18}} = -\frac{\sqrt{2}}{3}$$

$$\tan \tfrac{1}{2}\theta = -\sqrt{\frac{1 + \frac{5}{9}}{1 - \frac{5}{9}}} = -\sqrt{\frac{14}{4}} = -\frac{\sqrt{14}}{2}$$

EXAMPLE 3   Prove the identity

$$\cos^4 \theta = \tfrac{1}{8}(3 + 4 \cos 2\theta + \cos 4\theta)$$

Solution   The square of both members of formula (31) yields $\cos^2 \frac{1}{2}A = \frac{1}{2}(1 + \cos A)$. We employ this identity twice, first with $A = 2\theta$ and then with $A = 4\theta$, to obtain

$$\cos^4 \theta = (\cos^2 \theta)^2 = [\tfrac{1}{2}(1 + \cos 2\theta)]^2$$
$$= \tfrac{1}{4}(1 + 2 \cos 2\theta + \cos^2 2\theta)$$
$$= \tfrac{1}{4}[1 + 2 \cos 2\theta + \tfrac{1}{2}(1 + \cos 4\theta)]$$
$$= \tfrac{1}{8}(3 + 4 \cos 2\theta + \cos 4\theta)$$

Exercise 4-6

1. Find the values of $\sin 180°$ and $\cos 180°$ from functions of $90°$.
2. Find the values of the sine, cosine, and tangent of $60°$ from functions of $30°$.

Express each of the following as a function of twice the given angle.

3. $2 \sin 28° \cos 28°$ 4. $2 \sin 44° \cos 44°$
5. $\cos^2 48° - \sin^2 48°$ 6. $1 - 2 \sin^2 61°$
7. $2 \cos^2 40° - 1$ 8. $\sin 80° \cos 80°$
9. $\cos^2 \frac{1}{2}x - \frac{1}{2}$ 10. $\frac{1}{2} - \sin^2 3x$

In problems 11 through 14 find the values of the sine, cosine, and tangent of $2A$ and the quadrant in which $2A$ terminates.

11. A in Q I, $\sin A = \frac{12}{13}$ 12. A in Q II, $\tan A = -\frac{4}{3}$
13. A in Q III, $\cos A = -\frac{8}{17}$ 14. A in Q IV, $\sin A = -\frac{1}{3}$

Use the half-angle formulas to find the exact values of the sine, cosine, and tangent of each of the following angles.

15. $75°$ **16.** $15°$ **17.** $22\frac{1}{2}°$

18. $67\frac{1}{2}°$ **19.** $\downarrow 57\frac{1}{2}°$ **20.** $112\frac{1}{2}°$

Express each of the following as a function of half the given angle.

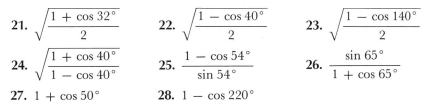

21. $\sqrt{\dfrac{1 + \cos 32°}{2}}$ **22.** $\sqrt{\dfrac{1 - \cos 40°}{2}}$ **23.** $\sqrt{\dfrac{1 - \cos 140°}{2}}$

24. $\sqrt{\dfrac{1 + \cos 40°}{1 - \cos 40°}}$ **25.** $\dfrac{1 - \cos 54°}{\sin 54°}$ **26.** $\dfrac{\sin 65°}{1 + \cos 65°}$

27. $1 + \cos 50°$ **28.** $1 - \cos 220°$

In problems 29 through 34 find the values of the sine, cosine, and tangent $\frac{1}{2}\theta$ if θ is between $0°$ and $360°$.

29. $\cos\theta = -\frac{4}{5},\ \theta$ in Q II **30.** $\sin\theta = -\frac{7}{25},\ \theta$ in Q IV

31. $\cos\theta = -\frac{2}{3},\ \theta$ in Q III **32.** $\sec\theta = \frac{6}{5},\ \theta$ in Q I

33. $\tan\theta = \frac{1}{3},\ \theta$ in Q I **34.** $\cos\theta = -\frac{3}{4},\ \theta$ in Q II

Prove the following identities.

35. $\sec 2\theta = \dfrac{\sec^2\theta}{\sec^2\theta - 2\tan^2\theta}$ **36.** $\sec 2\theta = \dfrac{-1}{1 - 2\cos^2\theta}$

37. $\sin 2\theta = \dfrac{2}{\sec\theta\csc\theta}$ **38.** $\tan 2\theta = \dfrac{2\cot\theta}{\cot^2\theta - 1}$

39. $\dfrac{1 + \cos 2x}{\sin 2x} = \cot x$ **40.** $\tan 4x = \dfrac{2\tan 2x}{1 - \tan^2 2x}$

41. $\cos^2\frac{1}{2}\theta = \dfrac{1 + \cos\theta}{2}$ **42.** $\sin^2\frac{1}{2}\theta = \dfrac{1 - \cos\theta}{2}$

43. $\cot\frac{1}{2}\theta = \dfrac{1 + \cos\theta}{\sin\theta}$ **44.** $\csc^2\frac{1}{2}\theta = \dfrac{2}{1 - \cos\theta}$

45. $\sin^2\frac{1}{2}\theta = \dfrac{\sec\theta - 1}{2\sec\theta}$ **46.** $\cot^2\theta = \dfrac{1 + \cos 2\theta}{1 - \cos 2\theta}$

47. $\cos 4x = 8\cos^4 x - 8\cos^2 x + 1$

48. $1 - \cos^2 2\theta = 4\sin^2\theta\cos^2\theta$

49. $2\sin^2 x + \cos 2x = 1$

50. $\cos 3x = 4\cos^3 x - 3\cos x$

51. $\tan(\theta - 45°) + \tan(\theta + 45°) = 2\tan 2\theta$

52. $\sin 4x = 4\sin x\cos x(2\cos^2 x - 1)$

53. $\tan 3\theta = \dfrac{\tan\theta(3 - \tan^2\theta)}{1 - 3\tan^2\theta}$

54. $\dfrac{1 + \sin 2x - \cos 2x}{1 + \sin 2x + \cos 2x} = \tan x$

55. $\sin^4 \theta = \frac{1}{8}(3 - 4\cos 2\theta + \cos 4\theta)$

56. $8\sin^2 \frac{1}{2}x \cos^2 \frac{1}{2}x = 1 - \cos 2x$

57. $\tan 4x = \dfrac{4\tan x(1 - \tan^2 x)}{1 - 6\tan^2 x + \tan^4 x}$

58. Multiply the numerator and denominator of the radicand in formula (32) by $1 - \cos A$ and simplify to obtain formula (33).

59. Derive formula (34).

4-9   The product and sum formulas

We shall next derive formulas which express the product of two functions as an algebraic sum of two functions and, conversely, the sum in terms of products. To make these derivations we rewrite formulas (16), (17), (23), and (24).

$$\sin A \cos B + \cos A \sin B = \sin (A + B)$$
$$\sin A \cos B - \cos A \sin B = \sin (A - B)$$
$$\cos A \cos B - \sin A \sin B = \cos (A + B)$$
$$\cos A \cos B + \sin A \sin B = \cos (A - B)$$

By forming the sum and difference of the members of the first two of these equations and the sum and difference of the last two, we get, respectively,

$$2\sin A \cos B = \sin (A + B) + \sin (A - B) \tag{35}$$
$$2\cos A \sin B = \sin (A + B) - \sin (A - B) \tag{36}$$
$$2\cos A \cos B = \cos (A + B) + \cos (A - B) \tag{37}$$
$$2\sin A \sin B = -\cos (A + B) + \cos (A - B) \tag{38}$$

These identities are called the *product formulas*. They express the product of a sine and a cosine, the product of two cosines, and the product of two sines as a sum or difference of sines and cosines. In short, the formulas express the product of sines and cosines as algebraic sums of sines and cosines. There are situations, especially in the calculus, in which the formulas may be advantageously employed.

We can convert the relations (35) to (38) into other useful forms by letting

$$A + B = x \qquad \text{and} \qquad A - B = y$$

or, when solved for A and B,

$$A = \tfrac{1}{2}(x + y) \qquad \text{and} \qquad B = \tfrac{1}{2}(x - y)$$

Making these substitutions for $A + B$, $A - B$, A, and B, we obtain

$$\sin x + \sin y = 2 \sin \tfrac{1}{2}(x + y) \cos \tfrac{1}{2}(x - y) \qquad (39)$$

$$\sin x - \sin y = 2 \cos \tfrac{1}{2}(x + y) \sin \tfrac{1}{2}(x - y) \qquad (40)$$

$$\cos x + \cos y = 2 \cos \tfrac{1}{2}(x + y) \cos \tfrac{1}{2}(x - y) \qquad (41)$$

$$\cos x - \cos y = -2 \sin \tfrac{1}{2}(x + y) \sin \tfrac{1}{2}(x - y) \qquad (42)$$

These identities are called the *sum formulas*. They express algebraic sums of functions in terms of products.

EXAMPLE 1   Express the product $\sin 28°\cos 20°$ as the sum of two functions.

Solution   From formula (35), we have

$$2 \sin 28°\cos 20° = \sin (28° + 20°) + \sin (28° - 20°)$$
$$\sin 28°\cos 20° = \tfrac{1}{2}(\sin 48° + \sin 8°)$$

EXAMPLE 2   Express the algebraic sum $\sin 5x - \sin 3x$ as a product.

Solution   Using formula (40), we get at once

$$\sin 5x - \sin 3x = 2 \cos 4x \sin x$$

EXAMPLE 3   Prove the identity $\dfrac{\sin 5x + \sin x}{\cos 5x - \cos x} = -\cot 2x.$

Solution   Applying formula (39) to the numerator and formula (42) to the denominator, we obtain

$$\frac{\sin 5x + \sin x}{\cos 5x - \cos x} = \frac{2 \sin 3x \cos 2x}{-2 \sin 3x \sin 2x} = -\cot 2x$$

Exercise 4-7

Express each of the following products as an algebraic sum.

1. $2 \sin 40°\cos 22°$
2. $2 \sin 50°\cos 15°$
3. $\cos 44°\cos 26°$
4. $\sin 36°\sin 26°$
5. $2 \sin 2\theta \cos \theta$
6. $2 \cos 5\theta \cos 2\theta$
7. $\sin 39°\cos 47°$
8. $\sin 17°\cos 41°$
9. $\cos 5\theta \cos 3\theta$
10. $\sin 4\theta \sin \theta$

Express each algebraic sum as a product.

11. $\sin 22° + \sin 18°$
12. $\sin 51° - \sin 30°$
13. $\cos 20° + \cos 5°$
14. $\cos 50° - \cos 45°$

15. $\cos 5x - \cos 3x$ **16.** $\cos 4x + \cos 6x$
17. $\cos 3x - \cos 2x$ **18.** $\sin 3x + \sin 7x$
19. $\cos 35° - \cos 40°$ **20.** $\sin 31° - \sin 41°$

Use one of formulas (35) to (42) and find the value of each expression.

21. $\sin 15° \cos 75°$ **22.** $2 \cos 45° \cos 15°$
23. $\sin 52\frac{1}{2}° \cos 7\frac{1}{2}°$ **24.** $\sin 52\frac{1}{2}° \sin 7\frac{1}{2}°$
25. $\sin 75° - \sin 15°$ **26.** $\sin 75° + \sin 15°$

Prove the following identities.

27. $\dfrac{\sin 4x + \sin 2x}{\cos 4x + \cos 2x} = \tan 3x$ **28.** $\dfrac{\cos 6x - \cos 8x}{\sin 6x + \sin 8x} = \tan x$

29. $\dfrac{\sin 5x - \sin 3x}{\sin 5x + \sin 3x} = \dfrac{\cot 4x}{\cot x}$ **30.** $\dfrac{\cos 2x + \cos 4x}{\cos 2x - \cos 4x} = \dfrac{\cot 3x}{\tan x}$

31. $\dfrac{\cos 4x + \cos 3x + \cos 2x}{\sin 4x + \sin 3x + \sin 2x} = \cot 3x$

32. $\dfrac{\sin x - \sin 2x + \sin 3x}{\cos x - \cos 2x + \cos 3x} = \tan 2x$

33. $2 \sin \theta \cos 2\theta + \sin 5\theta - \sin 3\theta = 2 \sin 2\theta \cos 3\theta$
34. $\cos 7\theta + \cos \theta + \cos 5\theta + \cos 3\theta = 4 \cos \theta \cos 2\theta \cos 4\theta$

4-10 Summary of formulas

For convenient reference we list here identities (16), (17), and (23) to (42) of this chapter.
    Functions of the sum and difference of two angles:

$$\cos (A - B) = \cos A \cos B + \sin A \sin B \qquad (16)$$
$$\cos (A + B) = \cos A \cos B - \sin A \sin B \qquad (17)$$
$$\sin (A + B) = \sin A \cos B + \cos A \sin B \qquad (23)$$
$$\sin (A - B) = \sin A \cos B - \cos A \sin B \qquad (24)$$

$$\tan (A + B) = \frac{\tan A + \tan B}{1 - \tan A \tan B} \qquad (25)$$

$$\tan (A - B) = \frac{\tan A - \tan B}{1 + \tan A \tan B} \qquad (26)$$

The double-angle formulas:

$$\sin 2A = 2 \sin A \cos A \qquad (27)$$

$$\cos 2A = \cos^2 A - \sin^2 A$$
$$= 1 - 2 \sin^2 A$$
$$= 2 \cos^2 A - 1 \tag{28}$$

$$\tan 2A = \frac{2 \tan A}{1 - \tan^2 A} \tag{29}$$

The half-angle formulas:

$$\sin \tfrac{1}{2}A = \pm \sqrt{\frac{1 - \cos A}{2}} \tag{30}$$

$$\cos \tfrac{1}{2}A = \pm \sqrt{\frac{1 + \cos A}{2}} \tag{31}$$

$$\tan \tfrac{1}{2}A = \pm \sqrt{\frac{1 - \cos A}{1 + \cos A}} \tag{32}$$

$$\tan \tfrac{1}{2}A = \frac{1 - \cos A}{\sin A} \tag{33}$$

$$\tan \tfrac{1}{2}A = \frac{\sin A}{1 + \cos A} \tag{34}$$

The product formulas:

$$2 \sin A \cos B = \sin (A + B) + \sin (A - B) \tag{35}$$
$$2 \cos A \sin B = \sin (A + B) - \sin (A - B) \tag{36}$$
$$2 \cos A \cos B = \cos (A + B) + \cos (A - B) \tag{37}$$
$$2 \sin A \sin B = -\cos (A + B) + \cos (A - B) \tag{38}$$

The sum formulas:

$$\sin A + \sin B = 2 \sin \tfrac{1}{2}(A + B) \cos \tfrac{1}{2}(A - B) \tag{39}$$
$$\sin A - \sin B = 2 \cos \tfrac{1}{2}(A + B) \sin \tfrac{1}{2}(A - B) \tag{40}$$
$$\cos A + \cos B = 2 \cos \tfrac{1}{2}(A + B) \cos \tfrac{1}{2}(A - B) \tag{41}$$
$$\cos A - \cos B = -2 \sin \tfrac{1}{2}(A + B) \sin \tfrac{1}{2}(A - B) \tag{42}$$

Exercise 4-8

Prove the following identities.

1. $\sec x \sec y \sin (x + y) = \tan x + \tan y$

2. $2 \csc 2x \sin (x - y) = \cos y \sec x - \sin y \csc x$

3. $\cos (x + y) \cos (x - y) = \cos^2 x \cos^2 y - \sin^2 x \sin^2 y$

4. $\sin (x + y) \sin (x - y) = \sin^2 x \cos^2 y - \cos^2 x \sin^2 y$

5. $(\sin\theta + \cos\theta)^2 = 1 + \sin 2\theta$

6. $(\sec\theta - \csc\theta)^2 = \sec^2\theta + \csc^2\theta - 4\csc 2\theta$

7. $\tan\theta - \tan\frac{1}{2}\theta = 2\csc 2\theta - \csc\theta$

8. $(1 + \tan^2\theta)\cot^2\theta = \csc^2\theta$

9. $\tan\frac{1}{2}\theta + 2\sin^2\theta + \cos 2\theta = 1 - \cot\theta + \csc\theta$

10. $2\tan\theta\cot 2\theta + \tan\frac{1}{2}\theta \doteq 1 + \csc\theta - \cot\theta - \tan^2\theta$

11. $\dfrac{\sin 3x}{\sin 2x} + \dfrac{\cos 3x}{\cos 2x} = \dfrac{2\sin 5x}{\sin 4x}$

Note: We use formulas (23) and (27) to show that the left member of the given identity is equal to the right member.

$$\frac{\sin 3x}{\sin 2x} + \frac{\cos 3x}{\cos 2x} = \frac{\sin 3x\cos 2x + \cos 3x\sin 2x}{\sin 2x\cos 2x}$$

$$= \frac{2\sin 5x}{\sin 4x}$$

12. $\dfrac{\sin 4x}{\cos 2x} - \dfrac{\cos 4x}{\sin 2x} = -\dfrac{2\sin 6x}{\sin 4x}$

13. $\dfrac{\cos 2x}{\sin 3x} + \dfrac{\sin 2x}{\cos 3x} = \dfrac{2\cos x}{\sin 6x}$

14. $\cos^4\theta - \sin^4\theta = 1 - 2\sin^2\theta$

15. $\dfrac{\sin^2 2x}{2\cos^2 x} = 1 - \cos 2x$

16. $\sin 2x = \dfrac{2\cot x}{1 + \cot^2 x}$

17. $\cos 2x = \dfrac{\cot^2 x - 1}{\cot^2 x + 1}$

18. $\dfrac{1 - \cos x}{1 + \cos\frac{1}{2}x} = 2(1 - \cos\frac{1}{2}x)$

19. $\dfrac{1 + \cos 2x}{1 - \sin x} = 2(1 + \sin x)$

20. $\dfrac{3\sec x + 1}{3\sec x - 1} = \dfrac{3 + \cos x}{3 - \cos x}$

21. $\dfrac{\cos\frac{1}{2}x + \sin\frac{1}{2}x}{\cos\frac{1}{2}x - \sin\frac{1}{2}x} = \dfrac{1 + \sin x}{\cos x}$

Hint: Multiply the numerator and denominator of the first fraction by $\cos\frac{1}{2}x + \sin\frac{1}{2}x$ and obtain

$$\frac{\cos^2\frac{1}{2}x + \sin^2\frac{1}{2}x + 2\sin\frac{1}{2}x\cos\frac{1}{2}x}{\cos^2\frac{1}{2}x - \sin^2\frac{1}{2}x}$$

Now apply formulas (13), (27), and (28).

22. $\sin x + \tan x = \dfrac{\sin x\tan x}{\tan\frac{1}{2}x}$

23. $\csc 2x = \dfrac{1 + \tan^2 x}{2 \tan x}$

24. $\dfrac{\sin^3 \theta - \cos^3 \theta}{\sin \theta - \cos \theta} = \dfrac{2 + \sin 2\theta}{2}$

Hint: First factor the numerator which is the difference of two cubes.

25. $\dfrac{4 \cos^3 \theta - \cos 3\theta}{4 \sin^3 \theta + \sin 3\theta} = \cot \theta$

26. $\tan (x + y) + \tan (x - y) = \dfrac{2 \tan x \sec^2 y}{1 - \tan^2 x \tan^2 y}$

27. $\dfrac{4}{\cos \theta - \cos 3\theta} - \dfrac{4}{\sin \theta + \sin 3\theta} = \sec \theta \csc \theta(\csc \theta - \sec \theta)$

Hint: Start by applying formula (42) to $\cos \theta - \cos 3\theta$ and formula (39) to $\sin \theta + \sin 3\theta$.

28. $\dfrac{2}{\cos 6\theta - \cos 2\theta} + \dfrac{2}{\sin 6\theta - \sin 2\theta} = \csc 2\theta(\sec 4\theta - \csc 4\theta)$

Chapter 4 Review exercise

1. Given $\cos \theta = -\frac{1}{4}$ and θ in quadrant III, find the values of the other functions of θ by use of the fundamental identities.

Find the values of the other five functions of θ in each problem 2 and 3.

2. $\cos \theta = -\frac{3}{5}$, θ in Q II   **3.** $\sec \theta = \dfrac{3}{\sqrt{5}}$, θ in Q IV

Reduce the first expression to the second in each problem 4 and 5.

4. $\dfrac{1 - \sec^2 \theta}{1 - \sec \theta}$; $1 + \sec \theta$ **5.** $\dfrac{1 - \sin^2 \theta}{1 + \cot^2 \theta}$; $\dfrac{1}{\csc^2 \theta \sec^2 \theta}$

Prove identities 6 and 7.

6. $\dfrac{4 + \csc \theta}{4 - \csc \theta} = \dfrac{4 \sin \theta + 1}{4 \sin \theta - 1}$ **7.** $\dfrac{1 - \cos^2 \theta - \tan^2 \theta}{\sin^2 \theta} = \dfrac{-1}{\cot^2 \theta}$

8. If $\sin A = \frac{4}{5}$ and $\cos B = \frac{5}{13}$ with A in quadrant I and B in quadrant IV, find the values of $\cos (A - B)$ and $\cos (A + B)$.

9. If $\sin A = \frac{3}{5}$ and $\cos B = \frac{12}{13}$ with A in quadrant II and B in quadrant IV, find the values of $\sin (A + B)$ and $\sin (A - B)$.

10. If $\tan A = \frac{8}{15}$ and $\tan B = -\frac{5}{12}$ with A in quadrant I and B in quadrant IV, find the values of $\tan (A + B)$ and $\tan (A - B)$.

11. If $\sin A = -\frac{4}{5}$ and $\tan B = -3$ with A in quadrant IV and B in quadrant II, find the values of the sine, cosine, and tangent of $(A + B)$ and $(A - B)$. Give the quadrant in which $(A + B)$ terminates and the quadrant in which $(A - B)$ terminates.

In each problem 12 and 13 find the values of $\sin 2A$, $\cos 2A$, and $\tan 2A$. In which quadrant does $2A$ terminate?

12. A in Q II, $\cos A = -\frac{12}{13}$ **13.** A in Q IV, $\tan A = -\frac{3}{4}$

In each problem 14 and 15 find the values $\sin \frac{1}{2}\theta$, $\cos \frac{1}{2}\theta$, and $\tan \frac{1}{2}\theta$ if θ is between $0°$ and $360°$.

14. $\tan \theta = -\frac{4}{3}$, θ in Q II   **15.** $\sin \theta = -\frac{7}{25}$, θ in Q III

Express each of the following products as an algebraic sum.

16. $2 \sin 41° \cos 25°$ **17.** $2 \sin 40° \cos 20°$
18. $2 \sin \theta \cos 3\theta$ **19.** $2 \cos 5\theta \cos 4\theta$
20. $\cos 3\theta \cos 6\theta$ **21.** $\sin 2\theta \sin 3\theta$

Express each of the following algebraic sums as a product.

22. $\sin 28° + \sin 18°$ **23.** $\cos 6x - \cos 2x$
24. $\sin 4x - \sin 6x$ **25.** $\cos 25° + \cos 55°$

graphical representation of the trigonometric functions

5-1 Variation of the trigonometric functions

The values of the trigonometric functions of an angle can be quite simply represented by directed line segments. We use Figs. 5-1 to 5-4 to provide a scheme for the representation. Each figure has a unit circle and a tangent line at the point Q. The terminal side of angle θ cuts the circle at the point P and the tangent line at R. Then we have

$$\sin \theta = \mathbf{MP} \qquad \cos \theta = \mathbf{OM} \qquad \tan \theta = \mathbf{QR}$$

A mental picture of the values which $\sin \theta$, $\cos \theta$, and $\tan \theta$ take as θ increases from 0 to 2π may be formed by focusing the attention on the resulting line segments $\mathbf{MP}$, $\mathbf{OM}$, and $\mathbf{QR}$.

As θ increases from 0 to $\pi/2$, the directed distance $\mathbf{MP}$, or $\sin \theta$, increases from 0 to 1; as θ increases from $\pi/2$ to π, $\sin \theta$ decreases from 1 to 0. For quadrants III and IV, the directed distance $\mathbf{MP}$ is negative. Hence as θ

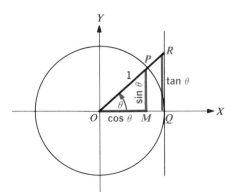

FIGURE 5-1

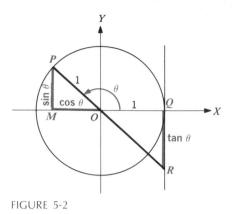

FIGURE 5-2

increases from π to $3\pi/2$, sin θ varies from 0 to -1. And finally, as θ increases from $3\pi/2$ to 2π, sin θ varies from -1 to 0.

In a similar way the variation of cos θ is revealed by the behavior of the line segment **OM.** For quadrants I and IV, the distance **OM** is positive and for quadrants II and III, the distance is negative. Consequently the variation of cos θ as θ increases from 0 to 2π is, in order of quadrants, 1 to 0, 0 to -1, -1 to 0, 0 to 1.

Figures 5-1 to 5-4 also enable us to visualize the variation of tan θ as θ increases. If θ starts at 0 and increases, the point R starts at Q and moves upward along the tangent line. The angle may be taken close enough to $\pi/2$ to make its tangent as large as any chosen positive number. For example, if the terminal side of θ passes through the point $(1,1000)$, tan θ is 1000. We conclude then that as θ increases from 0 to $\pi/2$, tan θ starts at 0 and increases through all positive values. The terminal side of $\pi/2$ does not intersect the tangent line. Hence there is no line representation of tan $\pi/2$. This agrees with an earlier observation that $\pi/2$ has no tangent (Sec. 2-4).

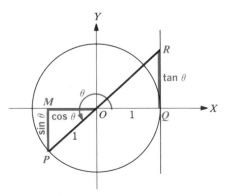

FIGURE 5-3

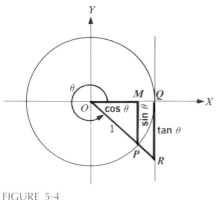

FIGURE 5-4

When θ is in the second quadrant, $\tan \theta$ is negative and we extend the terminal side of θ through the origin to obtain the line segment **QR.** This line segment, directed downward, is negative and has the proper length to represent $\tan \theta$. As θ increases from $\pi/2$ to π, the point R moves upward and reaches Q at $\theta = \pi$. Hence, in this process, $\tan \theta$ increases through all negative values to zero.

As θ increases from π to 2π, $\tan \theta$ repeats its behavior in quadrants III and IV exactly as in quadrants I and II, respectively.

The variations of $\sin \theta$, $\cos \theta$, and $\tan \theta$ are indicated in the following table. We suggest that the student prepare a similar table for the reciprocals $\cot \theta$, $\sec \theta$, and $\csc \theta$.

As θ varies from	$\sin \theta$ varies from	$\cos \theta$ varies from	$\tan \theta$ increases
0 to $\dfrac{\pi}{2}$	0 to 1	1 to 0	From 0 through all positive values
$\dfrac{\pi}{2}$ to π	1 to 0	0 to -1	Through all negative values to 0
π to $\dfrac{3\pi}{2}$	0 to -1	-1 to 0	From 0 through all positive values
$\dfrac{3\pi}{2}$ to 2π	-1 to 0	0 to 1	Through all negative values to 0

5-2   Periods of the trigonometric functions

A trigonometric function of an angle is not changed in value when the angle is changed by any integral multiple of 2π radians. For this reason the functions are said to be *periodic*. The sine function, for example, satisfies the equations

$$\sin \theta = \sin (\theta + 2\pi) = \sin (\theta + n \cdot 2\pi)$$

where θ is the radian measure of an angle and n is any integer. Thus the values of the sine of angles recur in intervals of 2π radians. The variation of the sine function is such that the recurrence does not take place in smaller intervals (Sec. 5-1), and consequently 2π radians is called the *period* of the function. The cosecant, cosine, and secant functions also have a period of 2π radians. To find the period of the tangent and cotangent functions, we set $A = \theta$ and $B = \pi$ in formula (25) of Sec. 4-6 and obtain

$$\tan (\theta + \pi) = \tan \theta$$

Hence the values of the tangent and cotangent of angles recur in intervals of π radians. Since the recurrence does not take place in a smaller interval, the period of these functions is π radians.

The property of periodicity makes the trigonometric functions useful in the study of numerous problems where phenomena are of a recurrent nature. As examples, the functions are employed in the study of sound waves, light waves, vibrating strings, the pendulum, and alternating currents and voltages.

An understanding of the periods of the trigonometric functions is helpful in constructing their graphs. If the graph of the tangent or cotangent function is obtained over an interval of π radians, then this part of the graph can be reproduced in other intervals of π radians. The same procedure is applicable to the other four functions except that the period is 2π radians.

Next, we proceed to discuss the graphs of the trigonometric functions. In this discussion, however, we shall let the domains of the functions be real numbers. In Sec. 2-5 we pointed out the close relation between the trigonometric functions of angles and the trigonometric functions of real numbers and discovered that either kind of domain seemed equally reasonable and logical. In this connection, we recall that the usual tables of trigonometric functions serve for finding the values of the functions of real numbers. Thus, illustrating with a real number x and the sine function, we have

$$\sin x = \sin (x \text{ radians}) = \sin \left(\frac{180x}{\pi} \right)^{\circ}$$

5-3   The graphs of the sine and cosine functions

To plot the graph of the sine function we use the equation $y = \sin x$. We first assign several real values to x in the interval 0 to 2π, choosing integral multiples of $\pi/6$. The sines of these numbers, equal to the sines of multiples of $30°$, may be found by referring to Table 2 in the Appendix. The values of x and the corresponding values of $\sin x$, or y, are shown in the table.

x	0	$\dfrac{\pi}{6}$	$\dfrac{\pi}{3}$	$\dfrac{\pi}{2}$	$\dfrac{2\pi}{3}$	$\dfrac{5\pi}{6}$	π
$y = \sin x$	0	0.50	0.87	1	0.87	0.50	0

x	$\dfrac{7\pi}{6}$	$\dfrac{4\pi}{3}$	$\dfrac{3\pi}{2}$	$\dfrac{5\pi}{3}$	$\dfrac{11\pi}{6}$	2π
$y = \sin x$	-0.50	-0.87	-1	-0.87	-0.50	0

We next lay off the same unit of length along the x axis and the y axis and take a point on the x axis about 3.14 units from the origin to correspond to the number π.† The corresponding values of x and y are plotted and a smooth curve is drawn through the points thus determined (Fig. 5-5). This part of the graph, which comprises one interval, may be duplicated in intervals of 2π to the right and left. The curve is called the *sine curve*.

By proceeding in just the same way as with the sine function, we can draw the graph of $y = \cos x$. The part of the cosine curve from $x = -\pi/2$ to $5\pi/2$ is shown in Fig. 5-6.

†The following plan, however, is reasonably accurate and easier to use. Choose a convenient number of spaces of the graph paper to stand for the length $\pi/3$ (equals 1.05 approximately) on the x axis and let the same number of spaces be the unit length on the y axis.

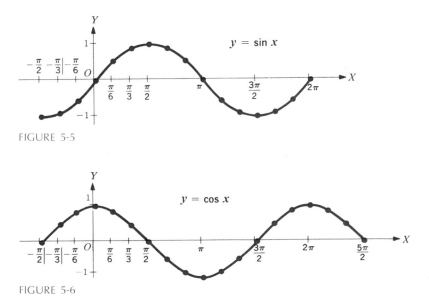

FIGURE 5-5

FIGURE 5-6

An examination of the sine curve and the cosine curve reveals that the values of these two functions vary from -1 to 1. These are called the *extreme values*. The extreme values and the zero values correspond to integral multiples of $\pi/2$. Moreover, the sine and cosine curves are shaped alike. Indeed, the identity $\sin(x + \frac{1}{2}\pi) = \cos x$ shows that the two curves drawn on the same coordinate axes would coincide if the sine curve were shifted $\frac{1}{2}\pi$ units to the left.

5-4   The graphs of the other functions

The period of the tangent function is π. Hence we may construct the graph of $y = \tan x$ for x between $-\frac{1}{2}\pi$ and $\frac{1}{2}\pi$ and then reproduce the curve in intervals of π to the right and left. The tabulated values for x and $\tan x$ may be verified by referring to Table 3 in the Appendix. To obtain the function values for negative values of x, we recall the identity $\tan(-x) = -\tan x$ [identity (22) of Sec. 4-5].

x	0	0.4	0.8	1	1.2	1.3
$\tan x$	0	0.42	1.0	1.6	2.6	3.6

The graph of the function (Fig. 5-7) is separated into *branches* at values of x which are odd integral multiples of $\pi/2$, where the tangent function does not exist. And the branches cross the x axis at integral multiples of π.

The graphs of the cotangent, secant, and cosecant functions are shown in Figs. 5-8 to 5-10. The graph of each function is separated into branches at values of x for which the function is not defined.

5-5   Graphs of more general functions

Having constructed the graphs of the trigonometric functions of the variable x, we next consider certain generalizations of the sine and cosine functions. Thus for the sine function we introduce the equations

$$y = a \sin bx \qquad \text{and} \qquad y = a \sin(bx + c)$$

where a, b, and c are constants. Since $\sin bx$ and $\sin(bx + c)$ vary from -1 to 1, the extreme values of the functions defined by the two equations are $-|a|$ and $|a|$. The period of $\sin bx$ may be obtained by determining how much x changes in producing a change of 2π in bx. As x varies from 0 to $2\pi/b$, the number bx varies from 0 to 2π if $b > 0$ and from 0 to -2π if $b < 0$. Accordingly, the period is $2\pi/b$. This is also the period of $\sin(bx + c)$.

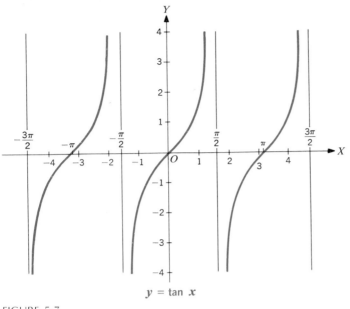

$y = \tan x$

FIGURE 5-7

The graphs of the equations $y = a \sin bx$ and $y = a \sin (bx + c)$, on the same coordinate axes, are alike except for position. This is true because the two values of y become equal if x is replaced by $x - c/b$ in the second equation. This tells us that a shift of c/b units along the x axis will bring the graph of the second equation into coincidence with the graph of the first

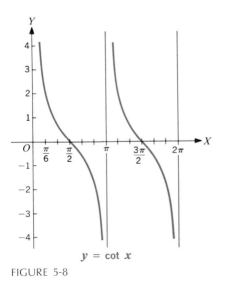

$y = \cot x$

FIGURE 5-8

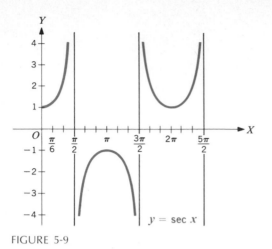

FIGURE 5-9

equation. The shift is to the right or left depending on whether b and c have
like or unlike signs.

Our discussion of the above equations applies in a similar way to the
equations $y = a \cos bx$ and $y = a \cos (bx + c)$. We proceed now to construct
graphs of special cases of the extended sine and cosine functions.

EXAMPLE 1   Draw the graph of the equation $y = \sin 3x$.

Solution   The function defined by this equation has extreme values of
-1 and 1 and a period of $2\pi/3$. Hence $\sin 3x$ and $\sin x$ have the same

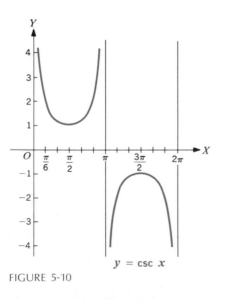

FIGURE 5-10

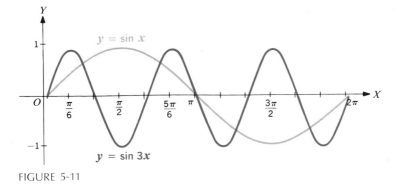

FIGURE 5-11

extreme values, but the period of sin $3x$ is one-third the period of sin x. These facts are exhibited in the graphs of $y = \sin x$ and $y = \sin 3x$ (Fig. 5-11).

EXAMPLE 2   Construct the graph of $y = 2 \cos \frac{1}{2}x$.

Solution   The extreme values are -2 and 2 and the period, 2π divided by $\frac{1}{2}$, is 4π. The graph constructed from the accompanying table of values is shown in Fig. 5-12. Each y value is twice the corresponding value of the graph of $y = \cos \frac{1}{2}x$, which is included for comparison.

x	0	$\frac{\pi}{2}$	π	$\frac{3\pi}{2}$	2π	$\frac{5\pi}{2}$	3π	$\frac{7\pi}{2}$	4π
$\frac{1}{2}x$	0	$\frac{\pi}{4}$	$\frac{\pi}{2}$	$\frac{3\pi}{4}$	π	$\frac{5\pi}{4}$	$\frac{3\pi}{2}$	$\frac{7\pi}{4}$	2π
$2 \cos \frac{1}{2}x$	2	1.4	0	-1.4	-2	-1.4	0	1.4	2

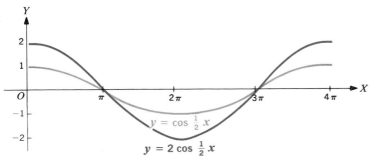

FIGURE 5-12

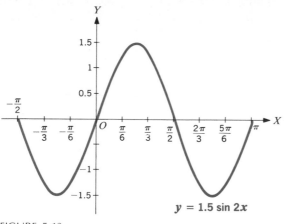

FIGURE 5-13

EXAMPLE 3   Draw the graph of the equation $y = 1.5 \sin 2x$.

Solution   The extreme values are -1.5 and 1.5, and the period is $2\pi/2 = \pi$. The graph (Fig. 5-13) resembles the graph of $y = \sin x$; comparatively, it is compressed horizontally and stretched vertically.

EXAMPLE 4   Construct the graphs of the equations

$$y = 1.5 \sin 2x \qquad \text{and} \qquad y = 1.5 \sin (2x + \tfrac{1}{3}\pi)$$

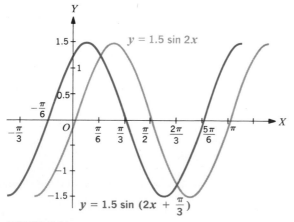

FIGURE 5-14

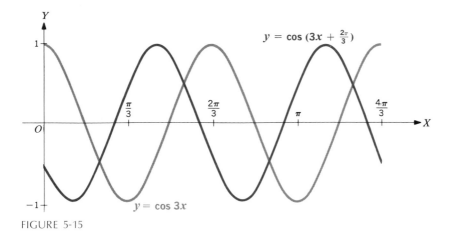

FIGURE 5-15

Solution   According to our discussion above, the graphs of the two equations are alike except for position (Fig. 5-14). And the curves would coincide if the graph of $y = 1.5 \sin (2x + \frac{1}{3}\pi)$ were moved $\frac{1}{6}\pi$ units to the right.

EXAMPLE 5   Draw the graphs of the equations

$$y = \cos 3x \qquad \text{and} \qquad y = \cos \left(3x + \frac{2\pi}{3}\right)$$

Solution   The graphs are drawn through two periods in Fig. 5-15. We note that the intersection points are at $x = 2\pi/9$, $5\pi/9$, $8\pi/9$, and $11\pi/9$. The graphs would coincide if the graph of the second equation were shifted $2\pi/9$ units to the right.

We remark that a rough drawing of each graph in Figs. 5-5 through 5-15 could be made without plotting points. The extreme values, the period, and a knowledge of the way in which the function varies suggest approximately where the curve should be drawn. This method of graphing is called *sketching*. In many problems a sketch serves as well as an accurately drawn curve.

5-6   Graphing by addition of ordinates

Frequently trigonometric functions of sums can be conveniently graphed by a process called *addition of ordinates*. As an illustration of this method, we shall construct the graph of the equation

$$y = 2 \cos x + \sin 3x$$

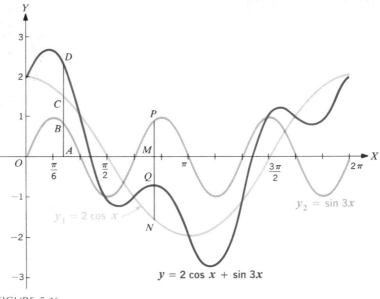

FIGURE 5-16

This plan requires that the graphs of $y_1 = 2 \cos x$ and $y_2 = \sin 3x$ be drawn on the same coordinate axes (Fig. 5-16). Then for any x the corresponding ordinate of $y = 2 \cos x + \sin 3x$ can be had from the two graphs. For example, the ordinate AD is the sum of the ordinates AB and AC. The addition of ordinates in this manner must be algebraic. Thus MN is negative, and the point Q is obtained by measuring downward from P so that $PQ = MN$. By plotting a sufficient number of points, in the way indicated, the desired graph can be drawn. This process of graphing is especially helpful when the graphs of the separate terms can be drawn easily.

Although the graph of an equation of the form

$$y = A \sin kx + B \cos kx$$

can be constructed by the addition of ordinates method, the appearance of kx in both terms of the right member opens the way for a more practicable procedure. Thus

$$A \sin kx + B \cos kx$$
$$= \sqrt{A^2 + B^2} \left(\frac{A}{\sqrt{A^2 + B^2}} \sin kx + \frac{B}{\sqrt{A^2 + B^2}} \cos kx \right)$$
$$= \sqrt{A^2 + B^2} \left(\sin kx \cos \theta + \cos kx \sin \theta \right)$$

where

$$\cos\theta = \frac{A}{\sqrt{A^2 + B^2}} \qquad \text{and} \qquad \sin\theta = \frac{B}{\sqrt{A^2 + B^2}}$$

We may now do the graphing by means of the equation

$$y = \sqrt{A^2 + B^2}\,\sin(kx + \theta)$$

EXAMPLE   Sketch the graph of the equation

$$y = 3\sin x - 4\cos x$$

Solution   Proceeding as above, we have

$$3\sin x - 4\cos x = 5\left(\tfrac{3}{5}\sin x - \tfrac{4}{5}\cos x\right)$$
$$= 5\left(\sin x \cos\theta + \cos x \sin\theta\right)$$

where $\cos\theta = 0.6$ and $\sin\theta = -0.8$. Referring to Table 3 in the Appendix, we find that θ is approximately equal to -0.93 for these values of the sine and cosine. Hence we obtain the graph of the given equation by using the equation

$$y = 5\sin(x - 0.93)$$

The graph of this equation (Fig. 5-17) is just the same as the graph of $y = 5\sin x$ moved 0.93 unit to the right.

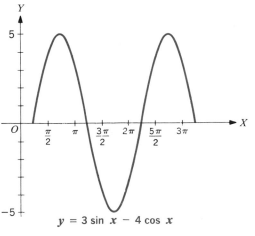

$$y = 3\sin x - 4\cos x$$

FIGURE 5-17

Exercise 5-1

Find the period of the graph of each equation. Give the extreme values of those involving the sine and cosine.

1. $y = \sin 4x$ **2.** $y = 3 \sin \frac{1}{2}x$ **3.** $y = -\sin \frac{1}{3}x$
4. $y = 4 \cos 6x$ **5.** $y = \frac{1}{2} \cos 5x$ **6.** $\cos \frac{2}{3}x$
7. $y = \cot 2x$ **8.** $y = \tan \frac{1}{3}x$ **9.** $y = \tan \frac{4}{3}x$
10. $y = \csc \frac{5}{3}x$ **11.** $y = \sec 8x$ **12.** $y = \csc 9x$

Construct one period of the graph of each equation. Start at $x = 0$.

13. $y = \cos 2x$ **14.** $y = -\cos 2x$ **15.** $y = 2 \sin 3x$
16. $y = 3 \sin 2x$ **17.** $y = 2 \cos \frac{1}{2}x$ **18.** $y = \frac{1}{2} \tan \frac{1}{2}x$

Sketch one period of the graph of each pair of equations on the same coordinate axes. Start at $x = 0$.

19. $y = 1.5 \sin x,\ y = 2 \cos x$ **20.** $y = \sin x,\ y = \sin 2x$

21. $y = \cos x,\ y = 2 \cos \frac{1}{2}x$ **22.** $y = \sin x,\ y = \sin\left(x + \dfrac{\pi}{3}\right)$

23. $y = \cos x,\ y = \cos\left(x - \dfrac{\pi}{6}\right)$

24. $y = \sin x,\ y = \sin\left(x + \dfrac{\pi}{4}\right)$

25. $y = 2 \cos x,\ y = \cos 3x$

Sketch one period of the graph of each equation by the method of addition of ordinates.

26. $y = \sin x - \cos x$ **27.** $y = \sin x - \cos 2x$
28. $y = \sin x + \sin 2x$ **29.** $y = 2 \cos x + \cos 3x$
30. $y = 2 \sin x + \sin 2x$ **31.** $y = 2 \cos x + \cos 2x$

Reduce each equation to the form $y = C \sin (kx + \theta)$ and sketch the graph.

32. $y = \sqrt{2}(\sin x - \cos x)$ **33.** $y = \sin x + \sqrt{3} \cos x$
34. $y = \sqrt{3}(\sin 3x + \cos 3x)$ **35.** $y = 3 \cos x + 4 \sin x$
36. $y = 2 \cos 2x - \sin 2x$ **37.** $y = \sin 2x + \cos 2x$

Chapter 5   Review exercise

Find the period of the graph of each equation. Give the extreme values of those involving the sine and cosine.

1. $y = 2 \cos x$ **2.** $y = 3 \sin \frac{1}{2}x$ **3.** $y = \cos \frac{2}{3}x$
4. $y = \tan \frac{1}{2}x$ **5.** $y = \sin 3x$ **6.** $y = \tan 4x$

Construct one period of the graph of each equation.

7. $y = \cos 2x$ **8.** $y = \sin 3x$ **9.** $y = \frac{1}{2} \tan x$

Sketch one period of the graph of each pair of equations on the same coordinate axes.

10. $y = \cos x, y = 2 \cos x$ **11.** $y = 2 \sin x, y = \sin \frac{1}{2}x$

Sketch one period of the graph of each equation by the method of addition of ordinates.

12. $y = \sin x + 2 \cos x$ **13.** $y = \cos x - \cos 2x$

GOVERNORS STATE UNIVERSITY
UNIVERSITY PARK
IL. 60466

trigonometric equations and inverse functions

6-1 Solutions of trigonometric equations

An equation involving one or more trigonometric functions of a variable is called a *trigonometric equation*. In Chap. 4 trigonometric equations of a particular kind, namely, trigonometric identities, were introduced. We shall now consider conditional trigonometric equations—equations which hold for only some of the angles (or numbers) for which the members are defined. Our problem will be to find solutions of such equations.

The simplest type of trigonometric equation is that in which a function of an angle is equal to a constant. Equations of this kind were met in finding unknown angles of a triangle (Chap. 3). The equation $\sin x = \frac{1}{2}$ is an example of a simple trigonometric equation. The equation is satisfied by $x = 30°$ and $x = 150°$ and all angles which differ from these by any integral multiple of $360°$. That is, for all integers n, the solutions are

$$x = 30° + n \cdot 360° \qquad \text{and} \qquad x = 150° + n \cdot 360°$$

When a trigonometric equation is not of the simplest type, the procedure is to derive two or more simple equations which yield all the solutions of the given equation. The tools for this procedure are algebraic operations and trigonometric identities.

At the outset, however, we note that the solutions of a trigonometric equation may be expressed by:

1. Angles in degree measure
2. Angles in radian measure
3. Real numbers (Sec. 2-5)

We expressed the solutions of $\sin x = \frac{1}{2}$ in degree measure. Henceforth, however, we shall express solutions in terms of real numbers unless it is necessary to use a table. When a table is needed in a problem, we shall refer to Table 2 and use degree measure. Also, for brevity, we shall give only the solutions in the interval $0 \leq x < 2\pi$ or in the interval $0° \leq x < 360°$.

EXAMPLE 1   Find all values of x in the interval $0 \leq x < 2\pi$ which satisfy the equation

$$2 \cos^2 x - 5 \cos x + 2 = 0$$

Solution   Factoring the left member gives

$$(\cos x - 2)(2 \cos x - 1) = 0$$

By setting each factor equal to zero, we obtain the simple equations

$$\cos x - 2 = 0 \quad , \text{and} \quad 2 \cos x - 1 = 0$$

The first of these equations has no solution because the value of a cosine never exceeds 1. From the second equation, we have $\cos x = \frac{1}{2}$, and the solutions are

$$x = \frac{\pi}{3} \quad \text{and} \quad x = \frac{5\pi}{3}$$

Each of these values may be changed by any integral multiple of 2π, but we are giving only the solutions in the interval $0 \leq x < 2\pi$. We check our results by substituting in the given equation. Thus

$$2 \cos^2 \frac{\pi}{3} - 5 \cos \frac{\pi}{3} + 2 = 2 \left(\frac{1}{2}\right)^2 - 5 \left(\frac{1}{2}\right) + 2 = 0$$

$$2 \cos^2 \frac{5\pi}{3} - 5 \cos \frac{5\pi}{3} + 2 = 2 \left(\frac{1}{2}\right)^2 - 5 \left(\frac{1}{2}\right) + 2 = 0$$

The two values satisfy the given equation and therefore the solution set is $\left\{\frac{\pi}{3}, \frac{5\pi}{3}\right\}$.

EXAMPLE 2   Find the solution set in the interval $0° \leq x < 360°$ of the equation

$$\tan^2 x - 3 \tan x - 2 = 0$$

Solution   This is a quadratic equation in $\tan x$, and applying the quadratic formula, we have

$$\tan x = \frac{3 \pm \sqrt{9 + 8}}{2} = \frac{3 \pm \sqrt{17}}{2} = 3.562, \ -0.5615$$

We use Table 2 and find a solution of $\tan x = 3.562$. Then the angle in the third quadrant having the solution thus found as a reference angle is also a solution. Similarly we find an angle whose tangent is $+0.5615$. The angles in the second and fourth quadrants having this angle as a reference angle are solutions of $\tan x = -0.5615$ (Sec. 3-4). Hence

$$\tan x = 3.562 \qquad\qquad \tan x = -0.5615$$
$$\phantom{\tan x = {}} x = 74°\ 19' \qquad\qquad x = 180° - 29°\ 19'$$
$$\phantom{\tan x = {}} = 150°\ 41'$$
$$\phantom{\tan x = {}} x = 180° + 74°\ 19' \qquad\qquad x = 360° - 29°\ 19'$$
$$\phantom{\tan x = {}} = 254°\ 19' \qquad\qquad = 330°\ 41'$$

These four results are solutions to the nearest minute. They check approximately in the given equation. Hence we express the solution set by

$$\{74°\ 19',\ 150°\ 41',\ 254°\ 19',\ 330°\ 41'\}$$

EXAMPLE 3   Find all values of x in the interval $0 \leq x < 2\pi$ which satisfy the equation

$$\sin 2x \sec^2 x = 2 \sin 2x$$

Solution   We express the equation with one member equal to zero and then factor the other member. Thus

$$\sin 2x \sec^2 x - 2 \sin 2x = 0$$
$$\sin 2x\, (\sec^2 x - 2) = 0$$

Equating the first factor to zero, we have

$$\sin 2x = 0$$
$$2x = 0,\ \pi,\ 2\pi,\ 3\pi$$
$$x = 0,\ \frac{\pi}{2},\ \pi,\ \frac{3\pi}{2}$$

From the second factor, $\sec^2 x - 2$, we obtain

$$\sec x = \sqrt{2} \qquad\qquad \sec x = -\sqrt{2}$$
$$x = \frac{\pi}{4},\ \frac{7\pi}{4} \qquad\qquad x = \frac{3\pi}{4},\ \frac{5\pi}{4}$$

All the values which we have found for x will satisfy the given equation except $\pi/2$ and $3\pi/2$. These values must be rejected because they have no secant. Hence the solution set is

$$\left\{0,\ \frac{\pi}{4},\ \frac{3\pi}{4},\ \pi,\ \frac{5\pi}{4},\ \frac{7\pi}{4}\right\}$$

We make two important observations concerning this problem. First, we did not divide the members of the given equation by $\sin 2x$. Such a division would have resulted in the loss of 0 and π as solutions. Second, in order to find the solutions of $\sin 2x = 0$ for $0 \leq x < 2\pi$ it is necessary to find the solutions for $0 \leq 2x < 4\pi$.

Exercise 6-1

Find the solution set of each equation in the interval $0 \le x < 2\pi$ or, when a table is used, the interval $0° \le x < 360°$.

1. $2 \sin x + 1 = 0$
2. $\cos x + 1 = 0$
3. $\tan x = 1$
4. $\cot x + \sqrt{3} = 0$
5. $2 \sin x + \sqrt{3} = 0$
6. $\sec x = 2$
7. $\csc x = -\sqrt{2} = 0$
8. $2 \cos x - \sqrt{3} = 0$
9. $\cos 2x = 0$
10. $\tan \frac{1}{2}x - 1 = 0$
11. $\sqrt{2} \sin 2x = 1$
12. $\sqrt{3} \sec 2x = 2$
13. $3 \sin \frac{1}{2}x - 1 = 0$
14. $\cot \frac{1}{2}x - 1 = 0$
15. $4 \cos \frac{1}{2}x = 1$
16. $\sin 3x = 0$
17. $2 \tan x - 3 = 0$
18. $4 \sin x = 3$
19. $4 \cos^2 x - 1 = 0$
20. $3 \cot^2 x - 1 = 0$
21. $4 \cos^2 x = 3$
22. $\cos^2 x + \cos x = 0$
23. $\sin x \cos x - \sin x = 0$
24. $\tan x - \cot x = 0$
25. $\sqrt{3} \csc x = 2 \sin x \csc x$
26. $\tan x \csc x - \sqrt{3} \csc x = 0$
27. $2 \sin^2 x + \sin x - 1 = 0$
28. $2 \cos^2 x + \cos x = 1$
29. $2 \cos^2 x + 5 \cos x + 2 = 0$
30. $3 \sin^2 x + 2 \sin x = 1$
31. $4 \sin^2 x + 3 \sin x = 1$
32. $\cos^2 x + \cos x - 6 = 0$
33. $2 \sec^2 x - \sec x = 1$
34. $3 \tan^2 x + 2 \tan x = 3$
35. $2 \cot^2 x - \cot x = 3$
36. $2 \csc^2 x + 3 \csc x = 2$
37. $\tan^2 x - 2 \tan x = 3$
38. $6 \sin^2 x + 5 \sin x = 6$
39. $3 \cos^2 x - 7 \cos x = 6$
40. $\tan^2 x + \tan x + 1 = 0$
41. $\sin^2 x + 5 \sin x + 2 = 0$
42. $\cos^2 x - 3 \sin x + 1 = 0$

6-2   Identities used in solving equations

We now consider trigonometric equations whose solutions can be found through use of certain of the identities (8) to (15) of Sec. 4-2 and the identities of Sec. 4-10. Here, as in the preceding section, the plan consists in finding simple equations (each containing one function of an angle or number) which have all the roots of the equation to be solved. When an operation is performed which might introduce extraneous roots, such as squaring both members of an equation or multiplying both members by a trigonometric expression involving the variable, it is essential that all results be tested in the given equation.

EXAMPLE 1   Solve the equation $\sin x = \cos 2x$.

Solution   This equation involves a function of x and a function of $2x$. But we obtain an equation containing a function of x only by applying identity

(28), $\cos 2x = 1 - 2 \sin^2 x$. Thus

$$\sin x = 1 - 2 \sin^2 x$$
$$2 \sin^2 x + \sin x - 1 = 0$$
$$(2 \sin x - 1)(\sin x + 1) = 0$$

The solution set, obtained by setting each factor equal to zero, is

$$\left\{ \frac{\pi}{6}, \frac{5\pi}{6}, \frac{3\pi}{2} \right\}$$

EXAMPLE 2   Solve the equation $2 \sin x + \cot x - \csc x = 0$.

Solution   First multiplying both members of the equation by $\sin x$, we get

$$2 \sin^2 x + \cos x - 1 = 0$$
$$2 (1 - \cos^2 x) + \cos x - 1 = 0$$
$$2 \cos^2 x - \cos x - 1 = 0$$
$$(2 \cos x + 1)(\cos x - 1) = 0$$

The first factor in the left member of the last equation yields $x = 2\pi/3$, $4\pi/3$, and the second factor yields $x = 0$. The values $2\pi/3$ and $4\pi/3$ will check in the given equation. But we reject $x = 0$ because neither the cotangent nor the cosecant is defined for this number. The extraneous root was introduced by multiplying the members of the given equation by $\sin x$.

EXAMPLE 3   Solve the equation $1 - \sin x = \cos x$.

Solution   To obtain an equation with only one function of x, we square the members of the given equation and substitute for $\cos^2 x$. Thus we get

$$1 - 2 \sin x + \sin^2 x = \cos^2 x$$
$$= 1 - \sin^2 x$$
$$2 \sin^2 x - 2 \sin x = 0$$
$$2 \sin x (\sin x - 1) = 0$$

The factors of the left side give $x = 0, \pi/2, \pi$. These results are solutions of the equation obtained by squaring the members of the original equation. Because this operation may have introduced extraneous roots, it is essential that each result be tested. Substituting in each side of the given equation, we find that

$x = 0$ gives $1 - 0 = 1$

$x = \dfrac{\pi}{2}$ gives $1 - 1 = 0$

$x = \pi$ gives $1 - 0 \neq -1$

We see that 0 and $\pi/2$ are roots but π is not a root. Hence the solution set is $\left\{ 0, \dfrac{\pi}{2} \right\}$.

EXAMPLE 4   Solve the equation $2 \sin 3x - 2 \sin x + 3 \cos 2x = 0$.

Solution   By identity (40), $\sin 3x - \sin x = 2 \cos 2x \sin x$. Then, from the given equation we get

$$4 \cos 2x \sin x + 3 \cos 2x = 0$$
$$\cos 2x \, (4 \sin x + 3) = 0$$

Setting the first factor equal to zero gives

$$\cos 2x = 0$$
$$2x = 90°, \ 270°, \ 450°, \ 630°$$
$$x = 45°, \ 135°, \ 225°, \ 315°$$

Equating the second factor to zero, we have $\sin x = -0.75$. Solutions of this equation have terminal sides in the third and fourth quadrants with the angle whose sine is $+0.75$ as a reference angle. The reference angle is $48° \ 35'$, and consequently we obtain

$$\sin x = -0.75$$
$$x = 180° + 48° \ 35' = 228° \ 35'$$
$$x = 360° - 48° \ 35' = 311° \ 25'$$

Combining the foregoing results, we have the solution set

$$\{45°, \ 135°, \ 225°, \ 315°, \ 228° \ 35', \ 311° \ 25'\}$$

Notice that we found all solutions of $\cos 2x = 0$ for $0° \leq 2x < 720°$. This enabled us to obtain all solutions of $\cos 2x = 0$ for which $0° \leq x < 360°$.

The operations which we have used in the preceding examples are often helpful in finding solutions of trigonometric equations. For emphasis, we list the operations in the form of directions.

1. If two or more angles (or numbers) appear in an equation, use the fundamental identities (8) to (15) of Sec. 4-2 and the identities of Sec. 4-10 to express the equation in terms of functions of a single angle (or number).
2. If different functions appear in the equation, use the fundamental identities to express the equation in terms of a single function.
3. Write the equation with one member equal to zero and the other member in factored form with each factor containing a single function, using the fundamental identities if necessary.
4. Simplify the equation by multiplying each member by a suitable trigonometric expression. Check to discover extraneous solutions.
5. Square both members of the equation. Check for extraneous solutions.

Exercise 6-2

Find the solution set of each equation in the interval $0 \leq x < 2\pi$ or, when a table is used, in the interval $0° \leq x < 360°$. Check the result when there is a possibility of an extraneous solution.

1. $\cot^2 x - \csc x - 1 = 0$
2. $2 \sin^2 x + \cos x + 4 = 0$
3. $\cos 2x - \sin x = 0$
4. $\sin \frac{1}{2}x - \cos x = 0$
5. $4 \cos x + \sec x - 4 = 0$
6. $\csc x + 4 \sin x + 4 = 0$
7. $\cos^2 x - \sin x + 5 = 0$
8. $\sec^2 x - \tan x - 3 = 0$
9. $\cot x + \csc x = -\sin x$
10. $\cot x + 2 \sin x = \csc x$
11. $\csc x + \sin x + 2 = 0$
12. $\tan x + \sec x = -\cos x$
13. $2 \cos x + 1 = \sin x$
14. $\cos x + \sqrt{3} \sin x = 1$
15. $\tan x + 2 \cos x = \sec x$
16. $1 - \sin x - \cos x = 0$
17. $\cos x - \sqrt{2} \sin x = 1$
18. $2 \cot x - \csc x = 0$
19. $3 \sin x - 2 = \cos x$
20. $2 \cos x - \sin x = 1$
21. $\cos 4x + \cos x = \cos 2x$
22. $\sin 3x - \sin x = 0$
23. $\sin 3x + \sin x = 0$
24. $\cos 3x - \cos x = 0$
25. $\cos^3 x \csc^3 x = 4 \cot^2 x$
26. $\sin^2 x + \sec^2 x = \cos^2 x$
27. $\sin 3x + \sin x = \sin 2x$
28. $\cot^3 - 8 = 0$
29. $\cot x + \csc x = 1$
30. $\tan 2x - \cos x = 0$
31. $\sin 2x + 2 \sin^2 \frac{1}{2}x = 1$
32. $2 \cos^2 \frac{1}{2}x + \sin 2x = 1$
33. $\sec x \tan x - \sqrt{2} = 0$
34. $\tan 2x + \tan x = 0$
35. $\tan \frac{1}{2}x + 2 \sin 2x = \csc x$
36. $\tan 2x + \cot x = 0$
37. $4 \sin^4 x - \cos^2 2x = 0$
38. $\sin^2 x \csc x + \csc x = 1$
39. $2 \cos^4 x - \sin^2 x = 0$
40. $3 \sin^4 x - 4 \cos^4 x = 0$
41. $2 \cot x + \csc x = 1$
42. $2 \tan x + \sec x = 1$
43. $\cos x + \sin x = \frac{1}{2}\sqrt{6}$
44. $\cos x - \sin x = \frac{1}{2}\sqrt{2}$

6-3 Inverse of a function

Let a function consist of the set of ordered pairs of numbers

$$f = \{(-5,7), (1,3), (2,-4), (8,10)\}$$

Now by interchanging the elements of each number pair of f, we obtain the set

$$g = \{(7,-5), (3,1), (-4,2), (10,8)\}$$

This new set of ordered pairs has one and only one second element corresponding to the first element of each ordered pair, and therefore is a function (Definition 1-5). The function g is said to be the *inverse* of the function f.

Next we interchange the first and second elements of each number pair of the function

$$s = \{(-3,6),\ (3,-4),\ (1,3),\ (8,-4)\}$$

and get

$$t = \{(6,-3),\ (-4,3),\ (3,\ 1),\ (-4,8)\}$$

The new set of number pairs here has two second elements, 3 and 8, corresponding to the first element -4. Hence the set t does not satisfy the definition of a function. The set, however, consists of ordered pairs of numbers and therefore is a relation.*

In the first example all the second elements of the ordered pairs of f are distinct, and consequently the interchange of the elements in each ordered pair yields a function. In the second example the second elements of the ordered pairs of t are not all distinct. This lack of uniqueness in the second elements caused the interchange of elements to yield a relation which is not a function. From these examples we see that the interchange of the elements of each ordered pair of a function will yield a function if and only if the second elements of the ordered pairs in the given function are all distinct. The new function obtained under this condition is said to be the inverse of the given function. If the given function is denoted by a letter, f say, then f^{-1} is the customary notation for the inverse function. We caution that -1, as used here, is *not* an exponent. We now state formally the definition of an inverse function.

Definition 6-1   *Let a function f be such that no two of its ordered pairs have the same second element. Then the inverse function f^{-1} is the set of ordered pairs obtained from f by interchanging the first and second elements of each ordered pair in f.*

We note that the domain and range of f become, respectively, the range and domain of f^{-1}. We note also that the interchange of the first and second elements of each ordered pair in f^{-1} will yield f. Hence f^{-1} is the inverse of f and f is the inverse of f^{-1}.

EXAMPLE 1   Find the inverse of the function determined by the equation

$$y = 2x - 3$$

where the domain is the set of all real numbers.

---

*A set of ordered pairs of numbers is called a *relation*. A function then is a relation, but, as this example illustrates, a relation is not necessarily a function.

Solution   The given equation, expressed in terms of x, becomes

$$x = \tfrac{1}{2}y + \tfrac{3}{2}$$

We obtain the inverse function by interchanging x and y in this equation. Thus

$$y = \tfrac{1}{2}x + \tfrac{3}{2}$$

EXAMPLE 2   Given the equation $y = x^2$, find a domain so that the resulting function will have an inverse.

Solution   The graph of the equation is a parabola opening upward as shown in Fig. 6-1. Clearly there are two y values corresponding to any x value and its negative. So let us restrict x to the set $\{x \geq 0\}$. The graph of the resulting function is the half of the parabola in the first quadrant. Now, solving the given equation for x and then interchanging the variables, we have

$$x = \pm\sqrt{y} \quad \text{and} \quad y = \pm\sqrt{x}$$

In order to obtain the inverse of the original function, we must reject negative values for y. Hence the inverse function is given by

$$y = \sqrt{x} \quad \text{for } x \geq 0$$

Figure 6-2 shows the graph of the original function and its inverse.

The student might draw the graphs of the functions defined by

$$y = x^2 \quad \text{for } x \leq 0$$

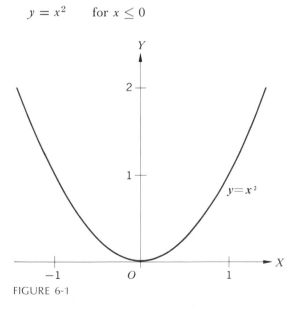

FIGURE 6-1

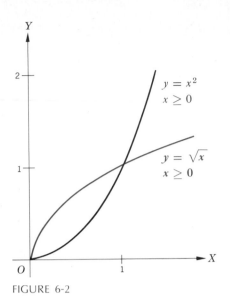

FIGURE 6-2

and

$$y = -\sqrt{x} \qquad \text{for } x \geq 0$$

and explain why the functions are inverses of each other.

6-4   The inverse trigonometric functions

Consider the equations $y = \sin x$ and $x = \sin y$. The graph of $y = \sin x$ winds along the x axis (Fig. 5-5). Clearly, then, the graph of $x = \sin y$ is of the same form but winds along the y axis (Fig. 6-3). If we assign x a value in the domain of $y = \sin x$, we obtain one definite corresponding value of y. If, on the other hand, we assign x any value from -1 to 1 in $x = \sin y$, we see that there are infinitely many corresponding values of y. These facts are exhibited in the graphs of the equations. Setting $x = 0$ in $x = \sin y$, for example, we have the resulting equation $\sin y = 0$. The solutions of this equation are $y = n\pi$ for all integral values of n.

   If we restrict the y values so that $-\pi/2 \leq y \leq \pi/2$, then the equation $x = \sin y$ has only one y value for each x value. To indicate this restriction on y, we use a capital S and write $x = \text{Sin}\, y$. The graph of this equation is the heavy part of the curve in Fig. 6-3. A y value in the equation $x = \text{Sin}\, y$ is the number whose sine is x. To indicate this relation of x and y, we use the notation $y = \text{Sin}^{-1} x$. Hence

$$y = \text{Sin}^{-1} x \qquad \text{is equivalent to} \qquad x = \text{Sin}\, y$$

Then either of these equations may be read "y is the number whose sine is x." We caution that $\mathrm{Sin}^{-1} x$ *does not mean* $1/\mathrm{Sin}\, x$.

The function defined by $y = \mathrm{Sin}^{-1} x$ is called an *inverse trigonometric function*, or, specifically, the *inverse sine function*. This designation comes from the way in which $y = \mathrm{Sin}^{-1} x$ is related to $y = \sin x$ with x restricted to the values $-\pi/2$ to $\pi/2$. We indicate this restriction on the domain of the sine function by using a capital S and writing $y = \mathrm{Sin}\, x$. Now any ordered pair (x,y) belonging to $y = \mathrm{Sin}\, x$ has the ordered pair (y,x) belonging to $y = \mathrm{Sin}^{-1} x$. This is true because the inverse sine function is obtained by first interchanging the variables in the equation $y = \sin x$. Consequently, the sets $\{-\pi/2 \le x \le \pi/2\}$ and $\{-1 \le y \le 1\}$ which constitute the domain and range of $y = \mathrm{Sin}\, x$ are interchanged to become the range and domain of $y = \mathrm{Sin}^{-1} x$.

Each of the other trigonometric functions has its inverse with notation similar to that for the inverse sine. Thus the inverse cosine, the inverse tangent, and the inverse cotangent functions are defined as follows:

$$y = \mathrm{Cos}^{-1} x \qquad \text{is equivalent to} \qquad x = \cos y,\ 0 \le y \le \pi$$

$$y = \mathrm{Tan}^{-1} x \qquad \text{is equivalent to} \qquad x = \tan y,\ -\pi/2 < y < \pi/2$$

$$y = \mathrm{Cot}^{-1} x \qquad \text{is equivalent to} \qquad x = \cot y,\ 0 < y < \pi$$

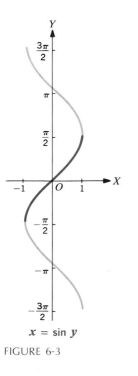

$x = \sin y$

FIGURE 6-3

The graphs of the four inverse functions which we have defined and the restricted trigonometric functions of which they are inverses are shown in Figs. 6-4 to 6-7. These inverse functions are used more frequently than are the inverse secant and the inverse cosecant functions. Consequently we shall not discuss the latter two functions.

For emphasis and convenience we list here the chosen domains and the resulting ranges of the four defined inverse functions.

Equation	Domain	Range
$y = \text{Sin}^{-1} x$	$-1 \le x \le 1$	$-\dfrac{\pi}{2} \le \text{Sin}^{-1} x \le \dfrac{\pi}{2}$
$y = \text{Cos}^{-1} x$	$-1 \le x \le 1$	$0 \le \text{Cos}^{-1} x \le \pi$
$y = \text{Tan}^{-1} x$	$-\infty < x < \infty$	$-\dfrac{\pi}{2} < \text{Tan}^{-1} x < \dfrac{\pi}{2}$
$y = \text{Cot}^{-1} x$	$-\infty < x < \infty$	$0 < \text{Cot}^{-1} x < \pi$

The notation for the inverse trigonometric functions, as written above, is widely used. There is, however, an equally acceptable notation in which Arcsin x is used instead of Sin^{-1}x, and similarly for the other functions.

EXAMPLE 1

$$\text{Sin}^{-1}\left(-\frac{1}{2}\right) = -\frac{\pi}{6} \qquad \text{Cos}^{-1}\left(\frac{-1}{\sqrt{2}}\right) = \frac{3\pi}{4}$$

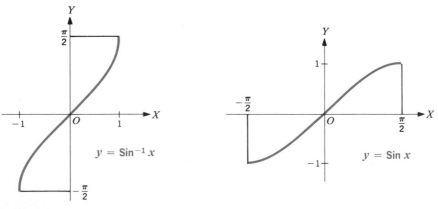

FIGURE 6-4

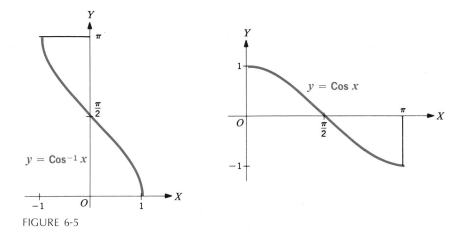

FIGURE 6-5

EXAMPLE 2

$$\text{Arctan } \sqrt{3} = \frac{\pi}{3} \qquad \text{Arccot } (-\sqrt{3}) = \frac{5\pi}{6}$$

The results shown in these examples, as we have indicated in our discussion of the inverse trigonometric functions, are real numbers. We could, however, interpret the results as angles expressed in radian measure. For example, $\text{Sin}^{-1}\left(-\frac{1}{2}\right)$ means the *number* whose sine is $-\frac{1}{2}$ or the *angle* whose sine is $-\frac{1}{2}$. Hence the result is the number $-\pi/6$ or the angle of measure $-\pi/6$ radian.

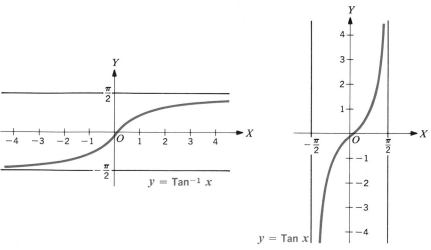

FIGURE 6-6

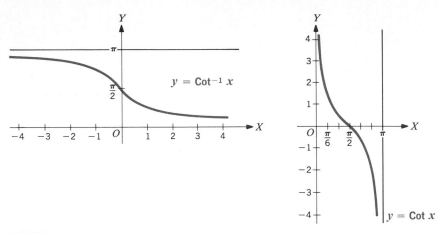

FIGURE 6-7

In some situations a trigonometric function of an inverse trigonometric function value is required. The following examples illustrate the procedure for handling a problem of this kind.

EXAMPLE 3   Find the value of $\tan [\text{Cos}^{-1}(-2/\sqrt{13})]$.

Solution   If we let $x = \text{Cos}^{-1}(-2/\sqrt{13})$, then by definition $\cos x = -2/\sqrt{13}$. Further, the range of the inverse cosine function requires that x have a value between 0 and π, and therefore

$$\sin x = +\sqrt{1 - \cos^2 x} = \sqrt{1 - \frac{4}{13}} = \frac{3}{\sqrt{13}}$$

Finally, using $\tan x = \sin x/\cos x$, we have

$$\tan \left[\text{Cos}^{-1} \left(-\frac{2}{\sqrt{13}} \right) \right] = -\frac{3}{2}$$

EXAMPLE 4   Find the value of $\sin [\text{Arctan}(-0.7291)]$.

Solution   Referring to Table 3, we find $\text{Arctan}(+0.7291) = 0.63$, and therefore $\text{Arctan}(-0.7291) = -0.63$. Then from the table, we find that

$$\sin [\text{Arctan}(-0.7291)] = \sin(-0.63) = -0.5891$$

EXAMPLE 5   Find the value of $\cos (2 \, \text{Sin}^{-1} \frac{3}{5})$.

Solution   Let $x = \text{Sin}^{-1} \frac{3}{5}$ or $\sin x = \frac{3}{5}$. We then need the value of $\cos 2x$. So, using the identity of $\cos 2x = 1 - 2 \sin^2 x$, we get

$$\cos (2 \, \text{Sin}^{-1} \tfrac{3}{5}) = 1 - 2(\tfrac{3}{5})^2 = 1 - \tfrac{18}{25} = \tfrac{7}{25}$$

EXAMPLE 6   Find the value of $\sin\left(\frac{1}{2}\,\mathrm{Cos}^{-1}\frac{2}{3}\right)$.

Solution   Let $x = \mathrm{Cos}^{-1}\frac{2}{3}$ or $\cos x = \frac{2}{3}$. We need the value of $\sin\frac{1}{2}x$. To get this value, we use the identity $\sin\frac{1}{2}x = \sqrt{(1 - \cos x)/2}$ and have

$$\sin\left(\frac{1}{2}\,\mathrm{Cos}^{-1}\frac{2}{3}\right) = \sqrt{\frac{1 - \frac{2}{3}}{2}} = \frac{\sqrt{6}}{6}$$

Exercise 6-3

Evaluate the following expressions. Use Table 3 when necessary.

1. $\mathrm{Sin}^{-1}\,0$
2. $\mathrm{Sin}^{-1}\,1$
3. $\mathrm{Cos}^{-1}\,0$
4. $\mathrm{Cos}^{-1}\,1$
5. $\mathrm{Tan}^{-1}\,0$
6. $\mathrm{Tan}^{-1}\,1$
7. $\mathrm{Sin}^{-1}\,(-1)$
8. $\mathrm{Cos}^{-1}\,(-1)$
9. $\mathrm{Tan}^{-1}\,(-1)$
10. $\mathrm{Cot}^{-1}\,\sqrt{3}$
11. $\mathrm{Sin}^{-1}\frac{1}{3}$
12. $\mathrm{Cos}^{-1}\frac{1}{2}$
13. $\mathrm{Cos}^{-1}\left(-\frac{\sqrt{3}}{2}\right)$
14. $\mathrm{Sin}^{-1}\left(-\frac{1}{\sqrt{2}}\right)$
15. $\mathrm{Cos}^{-1}\left(-\frac{\sqrt{2}}{2}\right)$
16. $\mathrm{Sec}^{-1}\,(-2)$
17. $\mathrm{Csc}^{-1}\,\sqrt{2}$
18. $\mathrm{Tan}^{-1}\,(-\sqrt{3})$
19. $\mathrm{Cos}^{-1}\,0.3986$
20. $\mathrm{Sin}^{-1}\,0.3153$
21. $\mathrm{Tan}^{-1}\,2.912$
22. $\mathrm{Sin}^{-1}\,(-0.8776)$
23. $\mathrm{Tan}^{-1}\,(-0.2027)$
24. $\mathrm{Cos}^{-1}\,(-0.9915)$
25. $\cos\left(\mathrm{Arcsin}\frac{1}{2}\right)$
26. $\tan\left[\mathrm{Arccos}\left(-\frac{1}{2}\right)\right]$
27. $\cot\left[\mathrm{Arcsin}\left(-\frac{1}{\sqrt{2}}\right)\right]$
28. $\sin\left[\mathrm{Arccos}\left(-\frac{1}{\sqrt{2}}\right)\right]$
29. $\tan\,(\mathrm{Arctan}\,\sqrt{2})$
30. $\sin\left[\mathrm{Arcsin}\left(-\frac{1}{3}\right)\right]$
31. $\cot\,[\mathrm{Arccot}\,(-3)]$
32. $\cos\left[\mathrm{Arccos}\left(-\frac{3}{4}\right)\right]$
33. $\mathrm{Arcsin}\,(\sin\frac{2}{3})$
34. $\mathrm{Arccos}\,(\cos\frac{1}{3})$
35. $\sin\,(2\,\mathrm{Arcsin}\frac{3}{4})$
36. $\cos\,(2\,\mathrm{Arccos}\frac{2}{3})$
37. $\tan\,(2\,\mathrm{Arccos}\frac{1}{3})$
38. $\cot\,(2\,\mathrm{Arcsin}\frac{2}{3})$
39. $\sin\left(\frac{1}{2}\,\mathrm{Cos}^{-1}\frac{1}{3}\right)$
40. $\mathrm{Cos}\left(\frac{1}{2}\,\mathrm{Sin}^{-1}\frac{3}{4}\right)$
41. $\tan\left(\frac{1}{2}\,\mathrm{Sin}^{-1}\frac{2}{3}\right)$
42. $\tan\left(\frac{1}{2}\,\mathrm{Cos}^{-1}\frac{1}{4}\right)$
43. $\sin\,[\mathrm{Cos}^{-1}\,(-0.7038)]$
44. $\cos\,[\mathrm{Sin}^{-1}\,(-0.3802)]$
45. $\tan\,[\mathrm{Sin}^{-1}\,(-0.9128)]$
46. $\tan\,[\mathrm{Cos}^{-1}\,(-0.6889)]$

6-5   Inverse trigonometric equations

In this section we consider certain expressions and equations involving inverse function values.

EXAMPLE 1   Verify the equality of the members of the equation

$$\mathrm{Cos}^{-1}\,\left(-\tfrac{12}{13}\right) - \mathrm{Sin}^{-1}\tfrac{4}{5} = \mathrm{Cos}^{-1}\,\left(-\tfrac{16}{65}\right)$$

Solution   We observe that the right member of the equation is a number between $\pi/2$ and π, and that its cosine is $-\frac{16}{65}$. So we test the validity of the equation by finding the cosine of the left member. Setting

$$A = \mathrm{Cos}^{-1}\left(-\tfrac{12}{13}\right) \qquad \text{and} \qquad B = \mathrm{Sin}^{-1}\tfrac{4}{5}$$

we have

$$\cos A = -\tfrac{12}{13} \qquad \sin A = \tfrac{5}{13} \qquad \sin B = \tfrac{4}{5} \qquad \cos B = \tfrac{3}{5}$$
$$\cos(A - B) = \cos A \cos B + \sin A \sin B$$
$$= \left(-\tfrac{12}{13}\right)\left(\tfrac{3}{5}\right) + \left(\tfrac{5}{13}\right)\left(\tfrac{4}{5}\right) = -\tfrac{16}{65}$$

We now see that the cosines of the members of the equation are equal, and consequently the equation is true. Showing that the sines of the members of the equation are equal would not definitely establish its validity. Explain why.

EXAMPLE 2   Solve the equation $\mathrm{Sin}^{-1} 3x + \mathrm{Sin}^{-1} x = \pi/2$.

Solution   We add $-\mathrm{Sin}^{-1} x$ to both members of the given equation and then take the cosine of the members of the resulting equation. Thus we get

$$\mathrm{Sin}^{-1} 3x = \frac{\pi}{2} - \mathrm{Sin}^{-1} x$$
$$\cos\left(\mathrm{Sin}^{-1} 3x\right) = \cos\left(\frac{\pi}{2} - \mathrm{Sin}^{-1} x\right)$$
$$\sqrt{1 - 9x^2} = \sin\left(\mathrm{Sin}^{-1} x\right)$$
$$1 - 9x^2 = x^2$$
$$x = \pm\frac{1}{\sqrt{10}}$$

It may be verified that $x = 1/\sqrt{10}$ satisfies the given equation. But the negative value for x must be rejected; it makes the left side of the given equation negative and therefore not equal to $\pi/2$.

EXAMPLE 3   Find the solution set of the equation

$$\cos\left(2\,\mathrm{Tan}^{-1} x\right) = \frac{1 - x^2}{1 + x^2}$$

Solution   First, we recall that the range of the inverse tangent function is $-\pi/2 < \mathrm{Tan}^{-1} x < \pi/2$. Then, checking, we see that both members of the given equation are positive when $-1 < x < 1$, and both members are negative when $x < -1$ and when $x > 1$. Both members are equal to zero when $x = \pm 1$. These results tell us that the members of the equation are defined for all real values of x. Further, the members do not differ in sign. So now we proceed to show that the members of the equation not only agree in sign but are also equal for all real values of x. For this purpose we use formula (28), Sec. 4-10, and the conditions $\cos (\text{Tan}^{-1} x) = 1/\sqrt{1 + x^2}$ and $\sin (\text{Tan}^{-1} x) = x/\sqrt{1 + x^2}$. Thus we get

$$\cos (2 \text{ Tan}^{-1} x) = \left(\frac{1}{\sqrt{1 + x^2}}\right)^2 - \left(\frac{x}{\sqrt{1 + x^2}}\right)^2$$

$$= \frac{1}{1 + x^2} - \frac{x^2}{1 + x^2}$$

$$= \frac{1 - x^2}{1 + x^2}$$

We now see that the members of the equation are equal for all real values of x. Hence the solution set may be expressed by $\{x \,|\, x \text{ is any real number}\}$.

EXAMPLE 4   Find the solution set of the equation

$$2 \text{ Cot}^{-1} x = \text{Sin}^{-1} \frac{2x}{x^2 + 1}$$

Solution   Recalling the ranges of the inverse cotangent function and the inverse sine function, we see that

1.   If x is a negative number or equal to zero, the left member of the equation is positive but the right member is not positive.
2.   If x is positive and less than 1, the left member exceeds $\pi/2$ but the right member cannot exceed $\pi/2$.
3.   If x is equal to 1, each member of the equation is equal to $\pi/2$.
4.   If x is a number greater than 1, the value of each member is between 0 and $\pi/2$.

And so we test the validity of the equation for values of x greater than 1 by taking the sines of the members of the equation. The sine of the right member is $2x/(x^2 + 1)$ and, for the left member, we find

$$\sin (2 \text{ Cot}^{-1} x) = 2 \sin (\text{Cot}^{-1} x) \cos (\text{Cot}^{-1} x)$$

$$= 2 \cdot \frac{1}{\sqrt{x^2 + 1}} \cdot \frac{x}{\sqrt{x^2 + 1}}$$

The sines of the members are equal, and therefore the solution set of the given equation is $\{x \geq 1\}$.

Exercise 6-4

Draw the graphs of the following equations.

1. $y = \text{Sin}^{-1} \frac{1}{2} x$ **2.** $y = \frac{1}{2} \text{Cos}^{-1} x$

3. $y = \text{Tan}^{-1} \frac{1}{2} x$ **4.** $y = 2 \text{Cot}^{-1} x$

Verify the following equations without using a table.

5. $\text{Tan}^{-1} 2 - \text{Tan}^{-1} 3 = \text{Tan}^{-1} \left(-\frac{1}{7} \right)$

6. $\text{Sin}^{-1} \frac{3}{5} + \text{Cos}^{-1} \frac{5}{13} = \text{Cos}^{-1} \left(-\frac{16}{65} \right)$

7. $\text{Tan}^{-1} \frac{1}{5} - \text{Tan}^{-1} \frac{2}{3} = \text{Tan}^{-1} \left(-\frac{7}{17} \right)$

8. $\text{Tan}^{-1} 7 + \text{Tan}^{-1} 2 = \text{Tan}^{-1} \left(-\frac{9}{13} \right)$

9. $\text{Tan}^{-1} 2 + \text{Tan}^{-1} 3 = \dfrac{3\pi}{4}$

10. $\text{Tan}^{-1} \frac{1}{7} + \text{Tan}^{-1} \frac{3}{4} = \dfrac{\pi}{4}$

Evaluate each of the following expressions.

11. $\cos \left(\text{Sin}^{-1} \frac{4}{5} - \text{Tan}^{-1} \frac{5}{12} \right)$ **12.** $\sin \left(\text{Cos}^{-1} \frac{5}{13} + \text{Tan}^{-1} \frac{12}{5} \right)$

13. $\tan \left(\text{Cos}^{-1} \frac{4}{5} + \text{Sin}^{-1} \frac{12}{13} \right)$ **14.** $\sin \left(\text{Cos}^{-1} \frac{1}{2} - \text{Sin}^{-1} \frac{3}{5} \right)$

15. $\cos \left[\left(\text{Sin}^{-1} \left(-\frac{3}{5} \right) - \text{Cos}^{-1} \left(-\frac{3}{5} \right) \right] \right.$ **16.** $\cos \left(\text{Sin}^{-1} x - \text{Cos}^{-1} x \right)$

17. $\sin \left(\text{Cos}^{-1} x + \text{Cos}^{-1} y \right)$ **18.** $\tan \left(\text{Tan}^{-1} x + \text{Tan}^{-1} 2y \right)$

19. $\cos \left(2 \text{Cos}^{-1} x + \text{Sin}^{-1} x \right)$ **20.** $\sin \left(2 \text{Tan}^{-1} x + \text{Cot}^{-1} x \right)$

Solve each of the following equations for x. Check the result.

21. $\text{Cos}^{-1} x - \text{Sin}^{-1} x = 0$ **22.** $\text{Tan}^{-1} x - \text{Cot}^{-1} x = 0$

23. $\text{Sin}^{-1} 2x - \text{Cos}^{-1} x = 0$ **24.** $\text{Sin}^{-1} 2\sqrt{x} = \text{Cos}^{-1} x$

25. $\text{Tan}^{-1} 2x + \text{Tan}^{-1} 3x = \dfrac{\pi}{4}$ **26.** $\text{Cos}^{-1} x + \text{Cos}^{-1} 2x = \dfrac{\pi}{2}$

27. $\text{Cos}^{-1} x + \text{Sin}^{-1} (1 - x) = \dfrac{\pi}{2}$ **28.** $\text{Sin}^{-1} \frac{1}{2} x = \text{Cos}^{-1} x$

29. $\text{Cos}^{-1} x = \text{Cos}^{-1} \dfrac{1}{2x}$ **30.** $\text{Sin}^{-1} x + \text{Cos}^{-1} (1 - x) = \dfrac{\pi}{2}$

Find the solution set of each equation.

31. $\cos \left(\text{Sin}^{-1} x \right) = \sqrt{1 - x^2}$ **32.** $\sin \left(\text{Cos}^{-1} x \right) = \sqrt{1 - x^2}$

33. $\tan \left(\text{Sin}^{-1} x \right) = \dfrac{x}{\sqrt{1 - x^2}}$ **34.** $\cos \left(\text{Tan}^{-1} x \right) = \dfrac{1}{\sqrt{1 + x^2}}$

35. $\cot \left(\text{Cos}^{-1} x \right) = \dfrac{x}{\sqrt{1 - x^2}}$ **36.** $\sin \left(\text{Tan}^{-1} x \right) = \dfrac{x}{\sqrt{1 + x^2}}$

37. $\cos\left(2\,\text{Cot}^{-1}x\right) = \dfrac{x^2 - 1}{x^2 + 1}$

38. $\cos\left(\dfrac{1}{2}\,\text{Cos}^{-1}x\right) = \sqrt{\dfrac{1 + x}{2}}$

39. $\sin\left(2\,\text{Tan}^{-1}x\right) = \dfrac{2x}{1 + x^2}$

40. $\text{Sin}^{-1}x + \text{Cos}^{-1}x = \text{Sin}^{-1}1$

41. $\text{Tan}^{-1}x + \text{Cot}^{-1}x = \text{Cos}^{-1}0$

42. $2\,\text{Sin}^{-1}x = \text{Cos}^{-1}\left(1 - 2x^2\right)$

43. $\text{Tan}^{-1}x + \text{Tan}^{-1}(-x) = 0$

44. $\dfrac{1}{2}\,\text{Cos}^{-1}x = \text{Cot}^{-1}\dfrac{\sqrt{1 - x^2}}{1 - x}$

45. $2\,\text{Tan}^{-1}x = \text{Cos}^{-1}\dfrac{1 - x^2}{1 + x^2}$

46. $2\,\text{Tan}^{-1}\dfrac{1}{x} = \text{Sin}^{-1}\dfrac{2x}{x^2 + 1}$

47. $2\,\text{Tan}^{-1}x = \text{Sin}^{-1}\dfrac{2x}{x^2 + 1}$

48. $2\,\text{Cot}^{-1}x = \text{Cos}^{-1}\dfrac{x^2 - 1}{x^2 + 1}$

Chapter 6   Review exercise

Find the solution set of each equation in the interval $0 \le x < 2\pi$ or, when a table is used, the interval $0° \le x < 360°$.

1. $\tan x - \sqrt{3} = 0$ **2.** $2\sin x - \sqrt{3} = 0$ **3.** $\sec x - 2 = 0$
4. $3\cos\frac{1}{2}x + 1 = 0$ **5.** $\tan\frac{1}{2}x + 1 = 0$ **6.** $4\sin\frac{1}{2}x + 1 = 0$
7. $2\sin^2 x - 5\sin x + 2 = 0$ **8.** $\tan^2 x + 3\tan x - 2 = 0$
9. $1 - \cos x = \sin x$ **10.** $\cos x = \sin 2x$
11. $2\cos x + \tan x - \sec x = 0$
12. $2\cos 3x - 2\cos x + 3\sin 2x = 0$

Evaluate the following expressions.

13. $\text{Sin}^{-1}0.4078$ **14.** $\text{Cos}^{-1}0.3058$ **15.** $\text{Tan}^{-1}2.820$
16. $\text{Sin}\left(\text{Tan}^{-1}\sqrt{2}\right)$ **17.** $\cos\left(\text{Sin}^{-1}\frac{1}{3}\right)$ **18.** $\left(\text{Tan}^{-1}\sqrt{2}\right)$

19. $\sin\left(\text{Sin}^{-1}\frac{1}{3}\right)$ **20.** $\tan\left[2\,\text{Sin}^{-1}\left(\dfrac{\sqrt{3}}{2}\right)\right]$ **21.** $\cos\left(\text{Cos}^{-1}0.1\right)$

22. $\cos\left(\text{Sin}^{-1}\frac{3}{5} + \text{Tan}^{-1}\frac{5}{12}\right)$ **23.** $\sin\left(\text{Cos}^{-1}\frac{12}{13} + \text{Tan}^{-1}\frac{12}{5}\right)$

logarithms

7-1 The logarithmic function

Logarithms are of much value in mathematics. By their use many numerical computations can be made with surprising simplicity. Of greater import, however, are the theoretical applications of logarithms in calculus and several other areas of more advanced mathematics. As we shall see, exponents lie at the basis of the study of logarithms. For this reason we recall from algebra certain definitions and laws of exponents.

1. $a^m a^n = a^{m+n}$

5. $\left(\dfrac{a}{b}\right)^m = \dfrac{a^m}{b^m}$

2. $\dfrac{a^m}{a^n} = a^{m-n}$

6. $a^0 = 1$

3. $(a^m)^n = a^{mn}$

7. $a^{-n} = \dfrac{1}{a^n}$

4. $(ab)^m = a^m b^m$

8. $a^{m/n} = \sqrt[n]{a^m}$

Definition 7-1   *If a is a positive number other than 1 and y is a real number in the equation $a^y = x$, then y is called the* logarithm *of x to the base a. That is, the logarithm of x is the exponent which the base a must have to yield a value equal to x.*

We write $y = \log_a x$ to mean that y is the logarithm of x to the base a. Hence the equations

$$a^y = x \qquad \text{and} \qquad y = \log_a x$$

express the same relation among the numbers $a, y,$ and x. The first equation is in *exponential form* and the second in *logarithmic form*.

In the definition of a logarithm we specify that the base $a > 0$ and that the exponent y is a real number (positive, negative, or zero). As a consequence, a^y is a positive number; hence we shall consider logarithms of positive numbers only. We assume (but do not prove) that if a and x are any positive numbers except $a = 1$, then there is one and only one real value of y which will satisfy the equation $a^y = x$. That is, every positive number has a unique logarithm with respect to a given base. We reject unity as a base because 1 with any exponent is still 1.

Since a positive number has a unique logarithm, the equation $y = \log_a x$ defines a function with the set of positive numbers as the domain and the set of real numbers as the range. Hence the function consists of the set of number pairs

$$\{(x, \log_a x) \,|\, x > 0\}$$

Suppose we choose $a = 3$ and construct the graph of the equation $y = \log_3 x$. We use the equivalent equation $x = 3^y$, assign integral values to y, and obtain the corresponding values of x as exhibited in the following table.

x	$\frac{1}{9}$	$\frac{1}{3}$	1	3	9
y	-2	-1	0	1	2

The graph (Fig. 7-1) is shown along with the graph of $y = \log_2 x$. Comparatively, we observe that $|\log_3 x| < |\log_2 x|$ except at $x = 1$. Both graphs cross the x axis at $x = 1$, as would be true for any base a because $a^0 = 1$.

As additional illustrations of the meaning of logarithms, we list some logarithmic equations and the corresponding equivalent exponential equations.

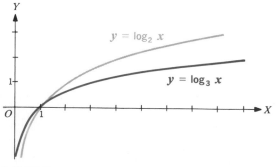

FIGURE 7-1

logarithmic form	exponential form
$\log_2 8 = 3$	$2^3 = 8$
$\log_9 27 = \frac{3}{2}$	$9^{3/2} = 27$
$\log_5 \frac{1}{25} = -2$	$5^{-2} = \frac{1}{25}$
$\log_a a = 1$	$a^1 = a$

EXAMPLE 1   Find x if $\log_5 x = -4$.

Solution   Using the equivalent exponential form, we have

$$x = 5^{-4} = \frac{1}{5^4} = \frac{1}{625}$$

EXAMPLE 2   Find a if $\log_a 36 = \frac{2}{3}$.

Solution   The exponential form is $a^{2/3} = 36$. We take the $\frac{3}{2}$ power of both members of this equation and obtain

$$(a^{2/3})^{3/2} = 36^{3/2} \qquad \text{and} \qquad a = 216$$

EXAMPLE 3   Find the value of $\log_7 \frac{1}{49}$.

Solution   We let $y = \log_7 \frac{1}{49}$. Then changing to the exponential form, we get

$$7^y = \frac{1}{49} = 7^{-2} \qquad \text{and} \qquad y = -2$$

Exercise 7-1

Give the logarithmic form of each of the following equations.

1. $2^6 = 64$
2. $5^3 = 125$
3. $4^{-1} = \frac{1}{4}$
4. $7^1 = 7$
5. $6^0 = 1$
6. $9^{3/2} = 27$
7. $81^{1/4} = 3$
8. $4^{-3/2} = \frac{1}{8}$
9. $8^2 = 64$
10. $(\frac{3}{2})^{-2} = \frac{4}{9}$
11. $(\frac{4}{3})^{-3} = \frac{27}{64}$
12. $(\frac{1}{25})^{-3/2} = 125$

Give the exponential form of each of the following equations.

13. $\log_3 1 = 0$
14. $\log_7 7 = 1$
15. $\log_4 16 = 2$
16. $\log_8 1 = 0$
17. $\log_2 8 = 3$
18. $\log_5 25 = 2$
19. $\log_{1/3} 9 = -2$
20. $\log_3 \frac{1}{27} = -3$
21. $\log_{1/4} 64 = -3$
22. $\log_{1/6} 36 = -2$
23. $\log_5 \frac{1}{25} = -2$
24. $\log_{3/2} \frac{81}{16} = 4$

Find x, a, or y in each of the equations.

25. $\log_{10} x = 2$
26. $\log_4 x = 1$
27. $\log_9 x = -1$
28. $\log_8 x = -\frac{2}{3}$
29. $\log_9 x = 0$
30. $\log_9 x = \frac{3}{2}$

31. $\log_{3/2} x = 4$ **32.** $\log_{9/4} x = \frac{3}{2}$ **33.** $\log_{1/3} x = -4$

34. $\log_a 16 = -\frac{2}{3}$ **35.** $\log_a 1 = 0$ **36.** $\log_a \frac{1}{16} = -4$

37. $\log_a \frac{1}{64} = -\frac{2}{3}$ **38.** $\log_a \frac{1}{4} = -1$ **39.** $\log_a \frac{9}{4} = -2$

40. $\log_a \frac{9}{16} = \frac{2}{3}$ **41.** $y = \log_3 9$ **42.** $y = \log_{10} 1000$

43. $y = \log_5 125$ **44.** $y = \log_{216} \frac{1}{6}$ **45.** $y = \log_4 \frac{1}{64}$

46. $y = \log_{25} 125$ **47.** $y = \log_{3/2} \frac{8}{27}$ **48.** $y = \log_{2/3} \frac{16}{81}$

7-2   Properties of logarithms

We shall state and prove three properties of logarithms. These properties follow readily from the laws of exponents.

Theorem 7-1   *The logarithm of a product is equal to the sum of the logarithms of the factors. That is,*

$$\log_a MN = \log_a M + \log_a N$$

Proof   Let $\log_a M = x$ and $\log_a N = y$. Then $M = a^x$, $N = a^y$,

$$MN = a^x a^y = a^{x+y} \qquad \text{and} \qquad \log_a MN = x + y$$

Substituting for x and y, we obtain

$$\log_a MN = \log_a M + \log_a N$$

For a product of three factors, as MNP, we have

$$\begin{aligned}
\log_a MNP &= \log_a (MN)(P) \\
&= \log_a MN + \log_a P \\
&= \log_a M + \log_a N + \log_a P
\end{aligned}$$

Similarly, the theorem may be extended to any number of factors.

EXAMPLE 1

$$\log_{10} (784)(92.7) = \log_{10} 784 + \log_{10} 92.7$$

Theorem 7-2   *The logarithm of a quotient is equal to the logarithm of the dividend minus the logarithm of the divisor. That is,*

$$\log_a \frac{M}{N} = \log_a M - \log_a N$$

Proof   If $\log_a M = x$ and $\log_a N = y$, then $M = a^x$ and $N = a^y$. Hence

$$\frac{M}{N} = \frac{a^x}{a^y} = a^{x-y} \qquad \text{and} \qquad \log_a \frac{M}{N} = x - y$$

By substituting for x and y, we obtain

$$\log_a \frac{M}{N} = \log_a M - \log_a N$$

EXAMPLE 2

$$\log_5 \tfrac{1078}{5234} = \log_5 1078 - \log_5 5234$$

Theorem 7-3 *The logarithm of a power of a number is equal to the exponent times the logarithm of the number. That is,*

$$\log_a M^p = p \log_a M$$

Proof We let $\log_a M = x$, or $M = a^x$. Then

$$M^p = (a^x)^p = a^{px} \quad \text{and} \quad \log_a M^p = px$$

Substituting for x gives

$$\log_a M^p = p \log_a M$$

EXAMPLE 3
$$\log_4 7842^3 = 3 \log_4 7842$$

EXAMPLE 4
$$\log_{10} 597^{-5} = -5 \log_{10} 597$$

EXAMPLE 5
$$\log_3 \sqrt{95} = \log_3 95^{1/2} = \tfrac{1}{2} \log_3 95$$

EXAMPLE 6 In this example we use Theorems 7-1 to 7-3 to express a logarithm as the algebraic sum of logarithms. Thus,

$$\log_2 \frac{28^2(6.7)^{1/2}}{(8.4)(17)} = \log_2 28^2(6.7)^{1/2} - \log_2 (8.4)(17)$$

$$= 2 \log_2 28 + \tfrac{1}{2} \log_2 6.7 - \log_2 8.4 - \log_2 17$$

EXAMPLE 7 Express $\log_6 5 - 3 \log_6 2 + \tfrac{2}{3} \log_6 28$ as a single logarithm.

Solution
$$\log_6 5 - 3 \log_6 2 + \tfrac{2}{3} \log_6 28 = \log_6 5 - \log_6 2^3 + \log_6 28^{2/3}$$

$$= \log_6 \frac{5(28)^{2/3}}{2^3}$$

Exercise 7-2

Write each of the following expressions as the algebraic sum of logarithms. A base is not indicated. Any positive number except 1 may be used.

1. $\log (152)(64)(81)$

2. $\log (43)^2(7.2)$

3. $\log (42)^{1/2}(53)^3$

4. $\log(38)^{1/3}(49)(8.51)$

5. $\log \dfrac{(4.9)(7.5)}{(56)(94)}$

6. $\log \dfrac{(7.2)^2(85)^{1/4}}{(66)(97)^2}$

7. $\sqrt{\dfrac{(83)(48)}{51}}$

8. $\log \sqrt{\dfrac{521}{(73)(55)}}$

9. $\log \dfrac{67 \sqrt[3]{38}}{40 \sqrt{12}}$

10. $\log \sqrt[4]{\dfrac{(68)(126)}{(79)(66)}}$

Write each expression as a single logarithm.

11. $\log 4 + \log \pi + \log r^3$

12. $\log x - \log y + \log z$

13. $\log \pi + 2 \log r + \log h$

14. $5 \log x - \log (2x - 3)$

15. $\log \sqrt{2x} - \frac{1}{2} \log (2x^2 - 3)$

16. $\frac{1}{3} \log 62 - \frac{3}{2} \log 4 - \log 5$

17. $\log 70 + 2 \log 726 - \log 91 - \frac{1}{2} \log 793$

18. $\frac{1}{2}(\log 21 + \log 34 + \log 52 - \log 877)$

Given $\log_{10} 2 = 0.3010$ and $\log_{10} 3 = 0.4771$, find the value of each of the following quantities.

19. $\log_{10} 12$

20. $\log_{10} 18$

21. $\log_{10} 27$

22. $\log_{10} \sqrt{48}$

23. $\log_{10} \sqrt[3]{36}$

24. $\log_{10} \sqrt[4]{54}$

25. $\log_{10} \frac{9}{64}$

26. $\log_{10} \frac{16}{27}$

27. $\log_{10} \frac{32}{81}$

7-3 Common logarithms

There are two systems of logarithms which are used extensively in mathematics. In one of these the base is an irrational number, designated by e, whose value is approximately 2.718. Logarithms to this base are called *natural*, or *Naperian*, *logarithms*. This system is convenient in many theoretical considerations and is used in calculus and higher mathematics. The other system employs 10 as a base. Logarithms to the base 10 are called *common*, or *Briggs*, *logarithms*. Common logarithms are especially suited for making computations.

We shall be interested in logarithms to the base 10; and henceforth, for brevity, this base will not be indicated in logarithmic expressions. Thus we shall write $\log_{10} N$ simply as $\log N$.

The common logarithms of the integral powers of 10 are integers. This is illustrated in the following list of powers of 10 and their logarithms.

$$10^3 = 1000 \qquad \log 1000 = 3$$
$$10^2 = 100 \qquad \log 100 = 2$$
$$10^1 = 10 \qquad \log 10 = 1$$
$$10^0 = 1 \qquad \log 1 = 0$$
$$10^{-1} = 0.1 \qquad \log 0.1 = -1$$
$$10^{-2} = 0.01 \qquad \log 0.01 = -2$$
$$10^{-3} = 0.001 \qquad \log 0.001 = -3$$

Noticing that $\log 1 = 0$, we may conclude that numbers greater than 1 have positive logarithms and that numbers between 0 and 1 have negative logarithms. Further, since the logarithm of an integral power of 10 is an integer, we conclude that the logarithms of other positive numbers are not integers. For example, 63 is between 10 and 10^2, and therefore its logarithm is between 1 and 2, or 1 plus a fraction (between 0 and 1). In general, then, a logarithm consists of two parts, an integer and a fraction.

7-4   The characteristic and mantissa

When a logarithm is expressed as an integer plus a decimal (between 0 and 1), the integer is called the *characteristic* and the decimal is called the *mantissa*. Finding the logarithm of a number, then, requires the determination of both the characteristic and the mantissa. We shall consider first the problem of finding the characteristic. Since $\log 1 = 0$ and $\log 10 = 1$, we assume (and it seems reasonable) that the logarithm of a number between 1 and 10 is zero plus a fraction between 0 and 1. In order to determine the characteristics corresponding to numbers between 0 and 1 and numbers greater than 10, we introduce the idea of the *reference position* of the decimal point.

Definition 7-2   *The reference position of the decimal point comes just after the first nonzero digit of the number.*

The arrow indicates the reference position of the decimal point of each of the following numbers.

$$3.741 \qquad 578.3 \qquad 0.456 \qquad 0.000125$$
$$\uparrow \qquad\quad \uparrow \qquad\quad\; \uparrow \qquad\qquad\; \uparrow$$

The decimal points of these numbers are respectively in the reference position, two places to the right of the reference position, one place to the left, and four places to the left.

Let us now discover how the characteristic of the logarithm of a number is related to the reference position of the decimal point of the number. To illustrate the procedure, we first determine the characteristic of the logarithm of each of the numbers 7832 and 0.07832. Expressing these numbers in scientific notation (Sec. 3-1), we have

$$7832 = 7.832 \times 10^3$$
$$0.07832 = 7.832 \times 10^{-2}$$

and, by Theorems 7-1 and 7-3,

$$\log 7832 = \log (7.832 \times 10^3) = 3 + \log 7.832$$
$$\log 0.07832 = \log (7.832 \times 10^{-2}) = -2 + \log 7.832$$

Thus we see that the characteristic of the logarithm of each of the numbers is the exponent of 10 when the number is expressed in scientific notation.

Consider now the general case in which a number A is expressed in scientific notation as $A = B \times 10^n$, where $1 \leq B < 10$ and n is an integer. Taking the logarithm of each member, we write

$$\log A = \log (B \times 10^n) = \log B + n$$

The mantissa of $\log A$ is equal to $\log B$ because $0 \leq \log B < 1$, and the characteristic is the integer n. If A is greater than 10, the characteristic is a positive integer; if A is between 0 and 1, the characteristic is a negative integer.

We note also that the characteristic of the logarithm of a number depends solely on the location of the decimal point. Further, to determine the characteristic, it is sufficient to notice what the exponent of 10 would be if the number were expressed in scientific notation. The mantissa, however, is independent of the position of the decimal point; it depends only on the sequence of digits in the number. Accordingly, we state the following.

Theorem 7-4   *If the decimal point of a number is in the reference position, the characteristic of its logarithm is zero. If the decimal point is n places to the right of the reference position, the characteristic is n. If the decimal point is n places to the left of the reference position, the characteristic is −n.*

The mantissa is independent of the position of the decimal point.

EXAMPLE 1   Find the characteristic of the logarithm of each of the numbers 0.0003405, 3.405, and 3405.

Solution   We count the number of places which the decimal point is
removed from the reference position in each case. Thus we find that the
characteristics are respectively -4, 0, and 3.

$$0.0003405 \qquad 3.405 \qquad 3405$$

$$-4 \qquad\qquad 0 \qquad\quad 3$$

EXAMPLE 2   Given log $5.826 = 0.7654$, write the logarithms of 582.6, $58{,}260$,
and 0.05826.

Solution   The mantissas of the logarithms of these numbers are all the
same as the given mantissa. Relative to the reference position, the decimal
points of 582.6, $58{,}260$, and 0.05826 are respectively two places to the right,
four places to the right, and two places to the left. Hence the required
characteristics are 2, 4, and -2, and we have

$$\log\ 582.6 = 2.7654$$
$$\log\ 58{,}260 = 4.7654$$
$$\log\ 0.05826 = -2 + 0.7654$$

We have noticed that the logarithm of a number between 0 and 1 is
negative. That is, the characteristic is a negative integer, and the mantissa
is either zero or a positive fraction. The characteristic may be combined with
the mantissa to yield a single negative quantity. For example,

$$\log 0.05826 = -2 + 0.7654 = -1.2346$$

For most computations, however, it is preferable to express a logarithm with
the fractional part positive.

The logarithm $-2 + 0.7654$ could not be written as -2.7654 since this
would indicate that the fractional part is also negative. A negative characteristic is customarily expressed as the difference of two integers. Thus, since
$-2 = 2 - 4$ and also equal to $8 - 10$, we have

$$\log 0.05826 = -2 + 0.7654$$
$$= 2.7654 - 4$$
$$= 8.7654 - 10$$

A negative characteristic may obviously be written as the difference of
two integers in an unlimited number of ways. The usual practice, however,
is to use a positive integer minus 10 or an integral multiple of 10.

7-5 Tables of logarithms

We have shown how to determine immediately the characteristic of the
logarithm of a number. Finding mantissas, however, is not simple. But

methods are developed in more advanced mathematics which permit the computation of a mantissa to any desired number of decimal places. The mantissas corresponding to many numbers have been computed and arranged in tabular form. Such an arrangement is called a *table of logarithms,* or a *table of mantissas.* Since most mantissas are unending decimal fractions, their values are approximated to two, three, or more decimal places. Accordingly a table is called a *two-place* table, a *three-place* table, and so on, depending on the number of decimal places used.

A four-place table (Table 4) is given in the Appendix. This table gives the mantissas of the numbers from 1.00 to 9.99 in steps of 0.01 of a unit. The first column, headed by N, contains the first two digits of a number and the third digit is to the right in the first line of the page. The mantissa corresponding to a number is in the horizontal line of the first two digits of the number and in the column headed by the third digit. As is customary, decimal points are omitted in the table. A decimal point belongs in front of each mantissa. Since the characteristic of the logarithms of 1 and numbers between 1 and 10 is zero, the table gives the logarithms of all three-digit numbers from 1.00 to 9.99. In order to use the table for a positive number outside this range, it is necessary to supply the proper nonzero characteristic in accordance with the discussion in Sec. 7-4. We shall explain with examples the use of the table in finding logarithms of numbers and in finding numbers corresponding to given logarithms.

EXAMPLE 1   Find (*a*) log 4.58, (*b*) log 0.00458.

Solution (*a*)   Since the decimal point is in the reference position, the characteristic of the logarithm is 0. To find the mantissa, look at Table 4 and go down the first column to 45, the first two digits in 4.58. Then look along the horizontal line from 45, stopping at the column which has 8 at the top, the third digit in 4.58. The mantissa .6609 is thus located. Hence

log 4.58 $=$ 0.6609

Solution (*b*)   The decimal point of 0.00458 is three places to the left of the reference position, and this makes the characteristic -3. Although the characteristic is determined from the position of the decimal point, the mantissa is independent of the decimal point. Hence the numbers 4.58 and 0.00458 have the same mantissa, and therefore

log 0.00458 $=$ 7.6609 $-$ 10

EXAMPLE 2   Find log 2047.

Solution   This number has four significant digits, and its mantissa cannot be read directly from the table. Hence we resort to interpolation, a process explained in Sec. 3-3. The number 2047 is 0.7 of the way from 2040 and

2050. The mantissas corresponding to 2040 and 2050 are the same as those for 204 and 205 and can be read directly from the table. The mantissas are 3096 and 3118, except for the decimal points. To obtain the number 0.7 of the way from the smaller to the larger mantissa, we compute 0.7(3118 − 3096). Our answer is 15, to the nearest integer, and hence 3096 + 15 = 3111 is the required mantissa when the decimal point is supplied. The quantities involved in this interpolation process are tabulated as a further aid in understanding the steps.

Number   Mantissa

$$10 \left[7 \begin{bmatrix} 2040 & 3096 \\ 2047 & ? \\ 2050 & 3118 \end{bmatrix} 22 \right]$$

$$0.7 \times 22 = 15.4$$
$$= 15 \quad \text{rounded off}$$
$$3096 + 15 = 3111$$

Since the required characteristic is 3, we have

$$\log 2047 = 3.3111$$

In beginning to find logarithms of numbers requiring interpolation the student may wish to tabulate the quantities involved as is done in this example. Having done a few problems, however, he should notice that most, if not all, of the operations can be done mentally. Emphasizing the process, we list the three essential steps.

1. A mantissa is to be subtracted from the one immediately following in the table.
2. This difference is to be multiplied by an integral multiple of tenths (0.1 or more up to 0.9).
3. The product, rounded off, is to be added to the smaller mantissa.

Exercise 7-3

Give the characteristic of the logarithm of each of the following numbers.

1. 84	**2.** 19	**3.** 824.2
4. 5.34	**5.** 200.5	**6.** 8.41
7. 50.45	**8.** 8745	**9.** 26,780
10. 0.273	**11.** 0.027	**12.** 0.006
13. 0.0009	**14.** 0.00002	**15.** 1.700
16. 5.75×10^{-1}	**17.** 87.2×10^{-2}	**18.** 85.6×10^{-2}
19. 2.30×10	**20.** 5.72×10^8	**21.** 2.04×10^{-4}

Find the common logarithm of each of the following numbers.

22. 31.6	**23.** 7.25	**24.** 2340
25. 47,200	**26.** 80.4	**27.** 0.123

28. 0.0082	**29.** 0.058	**30.** 0.236
31. 762	**32.** 800	**33.** 0.00087
34. 0.73	**35.** 657,000	**36.** 0.0571
37. 3.141	**38.** 46.44	**39.** 5729
40. 754.2	**41.** 0.8933	**42.** 0.001929
43. 48,320	**44.** 216,200	**45.** 0.007422
46. 28.55	**47.** 855.2	**48.** 809.6
49. 20.02	**50.** 0.2227	**51.** 201.4
52. 883.9	**53.** 0.000221	**54.** 0.02002
55. 0.000076	**56.** 456,400	**57.** 705.9
58. 4.339×10^5	**59.** 2.322×10^{-5}	**60.** 2.007×10^6
61. 8.104×10^{-9}	**62.** 8.103×10^{-7}	**63.** 8.863×10^8

7-6   Finding antilogarithms from tables

The number which corresponds to a logarithm is called the *antilogarithm* (abbreviated antilog). That is, if $\log N = x$, then $N = \text{antilog } x$. The following examples illustrate the use of Table 4 in finding antilogarithms.

EXAMPLE 1   Find antilog $(8.8407 - 10)$

Solution   We have the logarithm of a number and wish to find the number, or antilogarithm. Since the table has only the mantissas of logarithms, we disregard the characteristic for the moment and look in the columns of mantissas for 8407. The mantissas increase to the right along each row and from row to row. It is easy, therefore, to locate any mantissa which appears in the table. Having found the entry 8407, we look along its row to the left and find 69 in the N column. These are the first two digits of the antilogarithm. The third, found at the top of the column containing the given mantissa, is 3. Hence 693 is the sequence of digits corresponding to the mantissa .8407. To place the decimal point, we notice that the characteristic $8 - 10$ is equal to -2, and consequently the decimal point belongs two places to the left of the reference position. Thus

antilog $(8.8407 - 10) = 0.0693$

EXAMPLE 2   If $\log N = 4.7556$, find N.

Solution   The mantissa is not in the table, and we shall interpolate to find the antilogarithm. The mantissa 7556 is between the consecutive mantissas 7551 and 7559, and five-eighths of the way across. The numbers 569 and 570 or, equivalently, 5690 and 5700 correspond to the consecutive mantissas. Hence we find the number five-eighths of the way from 5690 to 5700. This gives 5696 which, except for the decimal point, is the required antilogarithm. Placing the decimal point in accordance with the given characteristic 4, we have

$N = 56{,}960$

The zero in this result serves to locate the decimal point; it is not a significant digit.

    The numbers involved in this example are presented in the following tabular form:

Number	Mantissa

$$10\begin{bmatrix}5690 & 7551 \\ ? & 7556 \\ 5700 & 7559\end{bmatrix}^{5}\Big]8 \qquad \begin{matrix}\tfrac{5}{8}\times 10 = 6 \quad \text{nearest integer} \\ 5690 + 6 = 5696 \\ N = 56{,}960\end{matrix}$$

Again, the student should observe that, in most cases, the computations required in finding antilogarithms can be done mentally.

Exercise 7-4

Use Table 4 and find the antilogarithm of each of the following logarithms.

1. 3.2148	**2.** 0.2718	**3.** 1.4997
4. 2.5224	**5.** 8.7825	**6.** 8.7118 − 10
7. 4.4330	**8.** 5.0253	**9.** 7.0212 − 10
10. 0.6248	**11.** 1.3468	**12.** 3.7155
13. 2.6120	**14.** 0.8016	**15.** 9.9710 − 10
16. 7.9937 − 10	**17.** 6.8719 − 10	**18.** 8.0923 − 10
19. 9.1139 − 10	**20.** 7.3355 − 10	**21.** 5.2520 − 10
22. 2.3229	**23.** 3.7157	**24.** 4.9819
25. 2.6047	**26.** 3.3434	**27.** 4.7240
28. 3.144	**29.** 6.5485	**30.** 5.2320

7-7   Logarithms used in computations

We are now ready to compute products, quotients, powers, and roots of numbers, or any combination of these operations, by means of logarithms. The logarithmic theorems of Sec. 7-2 serve for this purpose. We shall illustrate their use in some examples.

EXAMPLE 1   Compute N by means of logarithms, given

$$N = \frac{(32.41)(81.93)}{(1.854)(0.7949)}$$

Solution   Applying Theorem 7-2, we write

$$\log N = \log (32.41)(81.93) - \log (1.854)(0.7949)$$

And by Theorem 7-1

$$\log N = \log 32.41 + \log 81.93 - (\log 1.854 + \log 0.7949)$$

To help in avoiding errors and also to speed the work, the student should first make an outline form in which the characteristics of the given numbers are written and spaces are left for the quantities to be found. We suggest a form for this problem:

numerator   denominator

$$\begin{array}{rl} \log 32.41 = & 1 \\ (+)\ \log 81.93 = & 1 \\ \hline \log \text{num} = & \\ (-)\ \log \text{denom} = & \\ \hline \log N = & \end{array} \qquad \begin{array}{rll} \log 1.854 = & 0 \\ (+)\ \log 0.7949 = & 9 & -10 \\ \hline \log \text{denom} = & \end{array}$$

Having made the form, the tables should then be used and the spaces filled in. We show the completed form:

$$\begin{array}{rl} \log 32.41 = & 1.5106 \\ (+)\ \log 81.93 = & 1.9135 \\ \hline \log \text{num} = & 3.4241 \\ (-)\ \log \text{denom} = & 0.1684 \\ \hline \log N = & 3.2557 \\ N = & 1802 \end{array} \qquad \begin{array}{rll} \log 1.854 = & 0.2681 \\ (+)\ \log 0.7949 = & 9.9003 & -10 \\ \hline \log \text{denom} = & 10.1684 & -10 \\ = & 0.1684 \end{array}$$

EXAMPLE 2   Evaluate $\sqrt[3]{321.4}/\sqrt{208.7}$.

Solution   Letting N stand for the given fraction, we have

$$N = \frac{\sqrt[3]{321.4}}{\sqrt{208.7}} = \frac{(321.4)^{1/3}}{(208.7)^{1/2}}$$

and

$$\log N = \tfrac{1}{3} \log 321.4 - \tfrac{1}{2} \log 208.7$$

$$\log 321.4 = 2.5071 \qquad \log 208.7 = 2.3195$$

$$\tfrac{1}{3} \log 321.4 = 0.8357 \qquad \tfrac{1}{2} \log 208.7 = 1.1598$$

The logarithm of the denominator is greater than the logarithm of the numerator. We wish to obtain the difference, or $\log N$, in the form of a negative integer plus a positive fraction (mantissa). The table may then be used to find N. To obtain $\log N$ in this usable form, we first express the characteristic of the logarithm of the numerator as $10 - 10$. Thus we write

$$\tfrac{1}{3}\log 321.4 = 10.8357 - 10$$
$$(-)\,\tfrac{1}{2}\log 208.7 = \underline{\;\;1.1598\;\;}$$
$$\log N = \;\;9.6759 - 10$$
$$N = 0.4741$$

EXAMPLE 3   Compute $\sqrt[3]{0.8316}$.

Solution   We write $N = \sqrt[3]{0.8316} = (0.8316)^{1/3}$. Then

$$\log N = \tfrac{1}{3}\log 0.8316$$
$$\log 0.8316 = 9.9199 - 10$$

Dividing $9.9199 - 10$ by 3, we get the result, $3.3066 - 3.3333$, in an undesirable form. A usable form is obtained at once if the characteristic $9 - 10$ is replaced by the equal quantity $29 - 30$. Thus we get

$$\log 0.8316 = 29.9199 - 30$$
$$\tfrac{1}{3}\log 0.8316 = 9.9733 - 10$$
$$N = 0.9404$$

Exercise 7-5

Use logarithms to perform the indicated operations.

1. $(3.15)(0.738)$
2. $(2.38)(21.6)$
3. $33.4 \div 18.3$
4. $(327)(0.0094)$
5. $7.09 \div 65.7$
6. $422 \div 7.86$
7. $(18.5)(6.18)(64.7)$
8. $(4.02)(3.45)(9.66)$
9. $(92.4)(0.411)(8.47)$
10. $(552.3)(23.25)(707)$
11. $(36.21)(753.9)(0.761)$
12. $(7.102)(32.08)(1.873)$
13. $\dfrac{(5732)(4365)}{(8110)(3193)}$
14. $\dfrac{(4135)(7610)}{(455)(738)}$
15. $\dfrac{(3035)(9899)}{(7766)(1112)}$
16. $\dfrac{(2416)(6142)}{(7001)(6040)}$
17. $(2.204)^3$
18. $(0.0148)^4$
19. $\sqrt{8.041}$
20. $\sqrt[3]{4.004}$
21. $\sqrt[3]{57.31}$
22. $\sqrt[4]{0.0632}$
23. $(0.1776)^{2/3}$
24. $(21.08)^{3/4}$
25. $(80.11)^{-3/4}$
26. $(1.414)^{-5/2}$
27. $\dfrac{(5.635)^3}{\sqrt{1375}}$
28. $\dfrac{\sqrt{5625}}{\sqrt[3]{8140}}$

29. $\sqrt[3]{\dfrac{87.91}{(4851)(1584)}}$ **30.** $\sqrt[4]{\dfrac{(57.81)(6.183)}{91.87}}$

7-8   Exponential equations and logarithmic equations

An equation in which the unknown appears in an exponent is called an *exponential equation*. Some quite simple equations of this kind may be solved by inspection. Certain others are readily solved by the use of logarithms. The following examples illustrate methods of solving exponential equations.

EXAMPLE 1   Solve $5^x = 625^2$.

Solution   First express 625 as 5^4, then $5^x = (5^4)^2 = 5^8$. With the bases equal, the exponents must be the same; hence, $x = 8$.

EXAMPLE 2   Solve $2^x = 29$.

Solution   Since equal quantities have equal logarithms, we may write

$$\log 2^x = \log 29$$

Then

$$x \log 2 = \log 29$$

and

$$x = \frac{\log 29}{\log 2} = \frac{1.4624}{0.3010} = 4.858$$

The value of x is obtained by dividing log 29 by log 2. This is the quotient of two logarithms; it should not be confused with the logarithm of a quotient.

EXAMPLE 3   Solve the equation $5(3)^{2x+1} = 7^{1-x}$.

Solution   Taking the logarithm of each member gives

$$\log 5(3)^{2x+1} = \log 7^{1-x}$$

and

$$\log 5 + (2x + 1) \log 3 = (1 - x) \log 7$$

This is a linear equation in x and may be solved in the usual way. Thus

$$(2 \log 3 + \log 7)x = \log 7 - \log 5 - \log 3$$

and

$$x = \frac{\log 7 - \log 5 - \log 3}{2 \log 3 + \log 7} = \frac{-0.3310}{1.7993} = -0.1840$$

An equation that contains a logarithm of a variable is called a *logarithmic equation*. The following examples illustrate two types of such equations.

EXAMPLE 4   Find x if $\log (x - 2) + \log (x + 4) = \log 2$.

Solution   This equation may be expressed as $\log (x - 2)(x + 4) = \log 2$. Since quantities with equal logarithms are equal, we may write the equation $(x - 2)(x + 4) = 2$, whose solutions are $x = -1 + \sqrt{11}$ and $x = -1 - \sqrt{11}$. We reject the second of these values for x because $-1 - \sqrt{11}$ would give a logarithm of a negative number in the given operation.

EXAMPLE 5   Find x if $\log_4 (x + 3) + \log_4 (x - 1) = 2$.

Solution   This equation may be written as $\log_4 (x + 3)(x - 1) = 2$. Hence $(x + 3)(x - 1) = 4^2$ and $x^2 + 2x - 19 = 0$. So, we get $x = -1 + 2\sqrt{5}$ and $x = -1 - 2\sqrt{5}$. Tell why we must reject the second of these values for x.

Exercise 7-6

Solve problems 1 through 9 without the use of tables.

1. $4^x = 64$
2. $3^{x+1} = 9$
3. $e^{x-1} = e^8$
4. $4^{x-2} = 16^x$
5. $10^{3x-6} = 1000$
6. $16^x = 4$
7. $5^{x+2} = 25^x$
8. $5^x = \frac{1}{25}$
9. $5^{2x} = 5^0$

Solve the following exponential equations.

10. $3^x = 7$
11. $8^x = 15$
12. $(2.72)^x = 1.27$
13. $5^{2x-3} = 4$
14. $3^x = 5^x$
15. $(0.2)^x = 0.7$
16. $7(9)^x = 11^x$
17. $16^x = 3(17)^x$
18. $5(7)^x = 3(4)^x$
19. $2^{x-1} = 3^{x+2}$
20. $3(5)^{2x+1} = 4^{1-x}$
21. $(0.281)^x = 0.39$

Solve each of the following equations for x. Use 10 as the base of logarithms unless otherwise specified.

22. $\log (x - 3) + \log (x + 3) = \log 7$
23. $\log (x - 1) + \log (x + 3) = \log 5$
24. $\log (x - 4) + \log (x + 3) = \log 1$

25. $\log (x + 1) + \log (x - 2) = \log x$
26. $\log (4x + 3) - \log (x + 2) = \log x$
27. $\log (5 - x) + \log (3 - x) = \log 63$
28. $\log_5 (x - 2) + \log_5 (x + 2) = 2$
29. $\log_4 (x - 3) + \log_4 (x + 1) = 1$
30. $\log (x - 1) + \log (x + 3) = 1$
31. $\log (x - 1) + \log (x - 3) = 2$

7-9   Logarithms to different bases

It is sometimes desirable to express the logarithm of a number to one base in terms of its logarithm to another base. In particular, a change from a common logarithm to a natural logarithm, or vice versa, is advantageous in certain calculus problems. We shall derive the relation between the logarithms of a number to any two bases.

Let a and b stand for two positive numbers different from 1, and let

$$\log_a N = x \qquad \text{or} \qquad N = a^x$$

Taking the logarithm to the base b of both members of the exponential equation, we obtain

$$\log_b N = x \log_b a$$

and

$$x = \frac{\log_b N}{\log_b a}$$

Since $x = \log_a N$, we obtain

$$\log_a N = \frac{\log_b N}{\log_b a}$$

This formula can be used to change from one base to another. By substituting the natural base e (2.718 approximately) for a and 10 for b, we have

$$\log_e N = \frac{\log_{10} N}{\log_{10} e}$$

The values of $\log_{10} e$ and $1/\log_{10} e$, to four decimal places, are 0.4343 and 2.3026, respectively. Making these substitutions, we get the relations

$$\log_{10} N = 0.4343 \log_e N$$

and

$$\log_e N = 2.3026 \log_{10} N$$

EXAMPLES

$$\log_e 46 = 2.3026 \log_{10} 46 = (2.3026)(1.6628) = 3.829$$

$$\log_4 15 = \frac{\log_{10} 15}{\log_{10} 4} = \frac{1.1761}{0.6021} = 1.953$$

Exercise 7-7

Evaluate the following expressions.

1. $\log_e 60$ **2.** $\log_e 144$ **3.** $\log_e 0.408$
4. $\log_2 52$ **5.** $\log_3 160$ **6.** $\log_4 200$
7. $\log_5 0.061$ **8.** $\log_7 0.504$ **9.** $\log_{11} 140$
10. $\log_{12} 0.0184$ **11.** $\log_8 9.88$ **12.** $\log_9 0.007$

7-10   Logarithms of trigonometric functions

Table 5 in the Appendix gives to four decimal places the logarithms of the values of the sine, cosine, tangent, and cotangent of angles, in $10'$ steps from $0°$ to $90°$. The arrangement of the table is like that of Table 2. Because the values of the sine and cosine of angles between $0°$ and $90°$ are between 0 and 1, their logarithms are negative. The tangent of angles between $0°$ and $45°$ and the cotangent of angles between $45°$ and $90°$ also have negative logarithms. The characteristic of a negative logarithm is usually written as a positive integer minus 10. In Table 5, however, the -10 is omitted from every logarithmic entry and must be supplied by the user.

EXAMPLE 1   Find log tan $67° 27'$.

Solution   The angle $67° 27'$ is 0.7 of the way from $67° 20'$ to $67° 30'$. We assume this relation holds for their tangents. From the table we obtain the following figures:

$$\left.\begin{array}{l}\log \tan 67° 20' = 10.3792 - 10 \\ \log \tan 67° 30' = 10.3828 - 10\end{array}\right] 36$$

We add 0.7 of the tabular difference, 36, to the smaller logarithm. Thus $3792 + 0.7(36) = 3817$, and hence

$$\log \tan 67° 27' = 0.3817$$

EXAMPLE 2   Find log cos $15° 8'$.

Solution   Since the cosine decreases as the angle increases from $0°$ to $90°$, the correction must be subtracted in the interpolation process. We find log cos 15° = 9.9849 − 10

with 3 the tabular difference. Then 0.8 × 3 = 2.4 (2 rounded off) and therefore we get

log cos 15° 8′ = 9.9847 − 10

EXAMPLE 3   Find the angle A between 0° and 90° if log sin A = 9.7556 − 10.

Solution   The necessary computations follow:

$$10' \begin{bmatrix} \log \sin 34° \ 40' = 9.7550 \ -10 \\ \log \sin A \qquad = 9.7556 \ -10 \\ \log \sin 34° \ 50' = 9.7568 \ -10 \end{bmatrix} \begin{matrix} 6 \end{matrix} \Big] 18$$

Since $\frac{6}{18} \times 10' = 3'$, rounded off, we have $A = 34° \ 43'$.

Exercise 7-8

Find the value of the following quantities.

1. log sin 24° 10′
2. log cos 42° 20′
3. log tan 71° 40′
4. log cot 18° 30′
5. log sin 29° 14′
6. log cos 46° 42′
7. log cos 7° 8′
8. log sin 70° 32′
9. log tan 55° 3′
10. log cos 81° 22′
11. log sin 76° 52′
12. log cos 16° 36′
13. log cot 17° 7′
14. log tan 72° 4′
15. log tan 64° 13′
16. log cot 66° 29′

Find angle A between 0° and 90°.

17. log sin A = 9.9320 − 10
18. log cos A = 9.7549 − 10
19. log cos A = 9.8630 − 10
20. log sin A = 9.9371 − 10
21. log tan A = 9.8369 − 10
22. log cot A = 0.5103
23. log cot A = 9.9989 − 10
24. log tan A = 0.0068
25. log sin A = 9.6365 − 10
26. log cos A = 9.7465 − 10
27. log cot A = 9.1647 − 10
28. log tan A = 9.2183 − 10
29. log cos A = 9.9585 − 10
30. log sin A = 9.8202 − 10
31. log sin A = 9.4884 − 10
32. log cos A = 9.0150 − 10
33. log tan A = 0.0291
34. log cot A = 0.5461
35. log sin A = 9.4125 − 10
36. log cos A = 9.5925 − 10

Chapter 7   Review exercise

Give the logarithmic form of each equation.

1. $10^2 = 100$ **2.** $(\frac{3}{4})^3 = \frac{27}{64}$ **3.** $(\frac{1}{2})^4 = \frac{1}{16}$

Give the exponential form of each equation.

4. $\log_3 243 = 5$ **5.** $\log_2 1 = 0$ **6.** $\log_6 36 = 2$
7. $\log_5 25 = 2$

Write each expression as the algebraic sum of logarithms.

8. $\log_{10} \sqrt{(74)(83)}$ **9.** $\log_5 \frac{32}{64}$ **10.** $\log_{10} \dfrac{4\sqrt{2}}{3\sqrt{5}}$

Evaluate each logarithm, using interpolation.

11. $\log 0.3468$ **12.** $\log 26.81$ **13.** $\log 6574$

Find the value of each quantity. Use interpolation.

14. antilog 3.8283 **15.** antilog $9.8224 - 10$

Solve each equation for x.

16. $3^x = 81$ **17.** $3^{x-1} = 81$ **18.** $5^{x+1} = 125$
19. $\log(x - 4) + \log(x + 7) = \log 42$
20. $\log(x + 2) + \log(2x - 3) = 0$

8

oblique triangles

In Chap. 3 we discussed the solution of right triangles. We advance now to a consideration of oblique triangles—triangles with no right angle. Unlike the right triangle, the ratio of two sides of an oblique triangle does not in general yield a function of any of the angles. We shall, however, derive relations involving the sides and angles. These new relations, called *trigonometric laws,* will enable us to solve for the unknown parts of a triangle where there are sufficient known parts to determine the triangle. We shall take up the following cases of given parts of a triangle.

case I: A side and two angles
case II: Two sides and the angle opposite one of them
case III: Two sides and the included angle
case IV: Three sides

8-1   The law of sines

We begin our study of the oblique triangle by considering the case in which the length of a side and the measure of two angles are known.

Theorem 8-1   *In any triangle the sides a, b, c are proportional to the sines of the opposite angles A, B, C. That is,*

$$\frac{a}{\sin A} = \frac{b}{\sin B} = \frac{c}{\sin C}$$

Proof   In each of the triangles constructed in Figs. 8-1 and 8-2, h is an altitude drawn from the vertex of angle B. From either figure we have at once

$$\frac{h}{c} = \sin A \qquad \text{and} \qquad \frac{h}{a} = \sin C$$

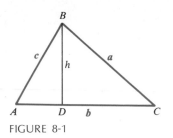

FIGURE 8-1

Solving each of these equations for h, we obtain

$$h = c \sin A \qquad \text{and} \qquad h = a \sin C$$

and, consequently,

$$a \sin C = c \sin A$$

Dividing by $\sin A \sin C$, we reduce the last equation to

$$\frac{a}{\sin A} = \frac{c}{\sin C}$$

By drawing the altitude from the vertex of C, we can show in the same way that

$$\frac{a}{\sin A} = \frac{b}{\sin B}$$

Combining the two results, we present the relation of the sides and angles in the form

$$\frac{a}{\sin A} = \frac{b}{\sin B} = \frac{c}{\sin C}$$

We see at once that the law of sines furnishes a method for solving case I. If two angles are known, the third angle can be found from the fact that the sum of the three angles is 180°. Then if b stands for the known side, the equations

$$\frac{a}{\sin A} = \frac{b}{\sin B} \qquad \text{and} \qquad \frac{c}{\sin C} = \frac{b}{\sin B}$$

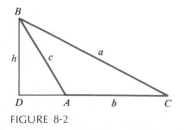

FIGURE 8-2

may be solved separately to yield a and c. If two sides and the angle opposite one of them are known, as a, b, and B, the equation

$$\frac{a}{\sin A} = \frac{b}{\sin B}$$

may be solved for its one unknown A.

We shall employ the law of sines first in solving triangles in which a side and two angles are known.

EXAMPLE 1   Solve the triangle which has $c = 231.7$, $A = 52°\ 17'$, and $B = 69°\ 53'$.

Solution   The triangle is drawn to scale in Fig. 8-3. Since the sum of the three angles is 180°, we have

$$C = 180° - (52°\ 17' + 69°\ 53') = 57°\ 50'$$

By the law of sines,

$$\frac{a}{\sin 52°\ 17'} = \frac{231.7}{\sin 57°\ 50'}$$

$$a = \frac{231.7 \sin 52°\ 17'}{\sin 57°\ 50'}$$

Hence

$$\log a = \log 231.7 + \log \sin 52°\ 17' - \log \sin 57°\ 50'$$

By use of the tables, we obtain

$$
\begin{aligned}
\log 231.7 &= 2.3649 \\
\log \sin 52°\ 17' &= 9.8982 - 10 \\
\hline
&= 12.2631 - 10 \\
\log \sin 57°\ 50' &= 9.9276 - 10 \\
\hline
\log a &= 2.3355 \\
a &= 216.5
\end{aligned}
$$

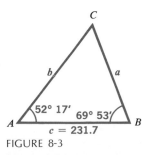

FIGURE 8-3

Using the law of sines again, we get

$$\frac{b}{\sin 69° \, 53'} = \frac{231.7}{\sin 57° \, 50'}$$

and

$$\log b = \log 231.7 + \log \sin 69° \, 53' - \log \sin 57° \, 50'$$

$$
\begin{aligned}
\log 231.7 &= 2.3649 \\
\log \sin 69° \, 53' &= 9.9727 - 10 \\
\hline
&12.3376 - 10 \\
\log \sin 57° \, 50' &= 9.9276 - 10 \\
\hline
\log b &= 2.4100 \\
b &= 257.1
\end{aligned}
$$

The computed parts are $C = 57° \, 50'$, $a = 216.5$, and $b = 257.1$.

EXAMPLE 2   A tree stands vertically on a hillside which makes an angle of 21° with the horizontal. From a point 50 feet directly down the hill from the base of the tree the angle of elevation of the top of the tree is 44°. Find the height of the tree.

Solution   We let a in Fig. 8-4 represent the height of the tree. Then from the triangle ABC, we have

$$A = 44° - 21° = 23°$$
$$B = 90° + 21° = 111°$$
$$C = 180° - (A + B) = 46°$$

Applying the law of sines, we obtain

$$\frac{a}{\sin 23°} = \frac{50}{\sin 46°}$$

$$a = \frac{50 \sin 23°}{\sin 46°}$$

$$= \frac{50(0.391)}{0.719} = 27 \text{ feet}$$

Exercise 8-1

Draw each triangle approximately to scale. Then solve for the unknown parts without the use of logarithms. Give the final results rounded off in accordance with the discussions in Secs. 3-1 and 3-7.

1. $c = 70$, $A = 44°$, $B = 65°$

2. $b = 30$, $A = 28°$, $C = 65°$

3. $b = 2.8$, $A = 33°$, $C = 72°$

4. $a = 8.2$, $B = 48°$, $C = 34°$

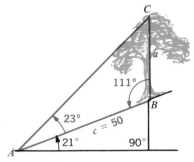

FIGURE 8-4

5. $a = 22$, $B = 25°$, $C = 95°$
6. $c = 50$, $B = 27°$, $C = 40°$
7. $b = 100$, $B = 52° \ 10'$, $C = 68° \ 00'$
8. $c = 300$, $B = 49° \ 30'$, $C = 83°20'$
9. $a = 242$, $A = 96° \ 00'$, $B = 34° \ 40'$

Use logarithms in computing the unknown parts of each triangle 10 through 15.

10. $b = 3.58$, $B = 68° \ 20'$, $C = 41° \ 50'$
11. $a = 347.4$, $B = 60° \ 22'$, $C = 81° \ 41'$
12. $b = 7632$, $A = 13° \ 20'$, $B = 40° \ 44'$
13. $c = 61.37$, $B = 80° \ 14'$, $C = 50° \ 12'$
14. $a = 700.1$, $B = 57° \ 1'$, $C = 64° \ 21'$
15. $b = 6514$, $A = 30° \ 3'$, $B = 63° \ 9'$

16. To find the distance between two points A and B on opposite sides of a river, we measure the distance from A to C to be 220 feet, the angle CAB to be $98° \ 40'$, and the angle ACB to be $41° \ 30'$. Compute the distance AB.
17. A tree stands vertically on a hillside which makes an angle of $21°$ with the horizontal. From a point 45 feet directly down the hill from the base of the tree the angle of elevation of the top of the tree is $41°$. Find the height of the tree.
18. Two men, 2100 feet apart, sight a balloon which is between them and in their vertical plane. The angle of elevation as measured by one of the men is $36°$ and by the other $52°$. Find the distance of the balloon from each observer and the height of the balloon.
19. Solve the preceding problem if the men are on the same side of the balloon, the other data unchanged.
20. A tree stands vertically on a hillside which makes an angle of $19°$ with the horizontal. From a point 170 feet directly up the hill from the tree the angle of depression of the top of the tree is $10°$. Find the height of the tree.

21. Two observers, 6 miles apart, see an object in the sky which is between them and in their vertical plane. The angle of elevation of the object from the position of one of the observers is 24° and from the position of the other is 32°. Find the altitude of the object. Next find the altitude if the observers are on the same side of the object.

8-2   The ambiguous case

As we have seen, a unique triangle is determined by a side and two angles. Case II, however, presents a new situation; there may be no triangle, one triangle, or two triangles, depending on the given parts. Because two triangles are sometimes constructible from two sides and the angle opposite one of them, case II is called the *ambiguous case*. The various possibilities for the ambiguous case are pictured in Figs. 8-5 through 8-10, where the given parts are designated by a, b, and A.

If $A \geq 90°$ and $a \leq b$, no triangle exists; a is too short to complete a triangle (Fig. 8-5). But if $a > b$, one triangle can be constructed (Fig. 8-6).

If A is an acute angle, we have the following possibilities:

1. If a is less than the perpendicular h, a triangle cannot be formed (Fig. 8-7).
2. If $a = h$, there is one triangle, a right triangle (Fig. 8-8).
3. If a is greater than h and less than b ($h < a < b$), two triangles are possible (Fig. 8-9).
4. If $a \geq b$, one triangle can be formed (Fig. 8-10).

The figures reveal that there is uncertainty only when A is acute and $a < b$. The decision in this instance can be made by applying the law of sines and computing $\sin B$. Thus,

$$\sin B = \frac{b \sin A}{a} = \frac{h}{a}$$

If the computed value for $\sin B > 1$, then $a < h$ and no triangle is possible (Fig. 8-7). If $\sin B = 1$, then $a = h$ and a right triangle can be formed (Fig.

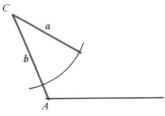

FIGURE 8-5

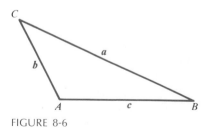

FIGURE 8-6

FIGURE 8-7   FIGURE 8-8

8-8). If $\sin B < 1$, then $a > h$ and two triangles are constructible (Fig. 8-9).

If logarithms are used in computing, $\log \sin B > 0$ means no triangle, $\log \sin B = 0$ means one triangle, and $\log \sin B < 0$ means two triangles.

EXAMPLE 1   Given $a = 17$, $b = 28$, and $B = 108°$, find angle A.

Solution   It is easy to see that there is only one solution. Applying the inverted form of the law of sines, which places the unknown in the numerator, we have

$$\frac{\sin A}{17} = \frac{\sin 108°}{28}$$

$$\sin A = \frac{17 \sin 108°}{28} = 0.577$$

$$A = 35° \qquad \text{nearest degree}$$

EXAMPLE 2   Determine if a triangle can be formed with the parts $a = 41$, $b = 62$, and $A = 43°$.

Solution   Substituting in the law of sines gives

$$\frac{\sin B}{62} = \frac{\sin 43°}{41}$$

$$\sin B = 1.03$$

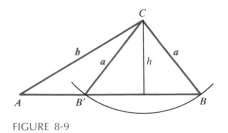

FIGURE 8-9

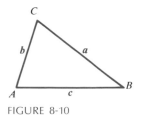

FIGURE 8-10

Since the sine of an angle can never exceed 1 we conclude that a triangle cannot be formed with the given parts. This situation is pictured in Fig. 8-7.

EXAMPLE 3   Solve the triangle whose given parts are $a = 182$, $b = 243$, and $A = 43° 20'$.

Solution   Since A is acute and $a < b$, there may be no, one, or two solutions. By constructing angle A and taking a distance along one side to represent b (Fig. 8-11), we surmise that a may be drawn in either of two positions to complete a triangle. And there are two solutions because, as we shall see, $\sin B < 1$. We solve first the triangle made by placing a so that angle B is acute (Fig. 8-12). We have

$$\frac{\sin B}{243} = \frac{\sin 43° 20'}{182}$$

$$\sin B = \frac{243 \sin 43° 20'}{182} = \frac{243(0.6862)}{182} = 0.9162$$

$$B = 66° 22'$$

$$C = 180° - (43° 20' + 66° 22') = 70° 18'$$

To find c, we write

$$\frac{c}{\sin 70° 18'} = \frac{182}{\sin 43° 20'}$$

$$c = \frac{182 \sin 70° 18'}{\sin 43° 20'} = \frac{182(0.9415)}{0.6862} = 249.7$$

Having found B, C, and c, we next solve the triangle formed by placing side a in the position shown in Fig. 8-13. We designate the unknown parts of this triangle by B', C', and c'. Here angle B' is obtuse, but $\sin B'$, expressed in terms of the given parts, is exactly the same as $\sin B$ for the first triangle. We conclude therefore that B' and B are supplementary.

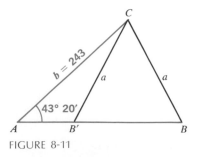

FIGURE 8-11

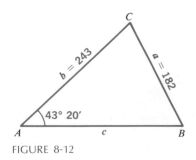

FIGURE 8-12

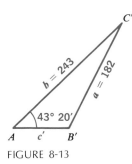

FIGURE 8-13

Hence B' and C' are readily obtainable. Thus

$B' = 180° - B = 180° - 66° 22' = 113° 38'$
$C' = 180° - (43° 20' + 113° 38') = 23° 2'$

The law of sines now yields c'.

$$\frac{c'}{\sin 23° 2'} = \frac{182}{\sin 43° 20'}$$

$$c' = \frac{182 \sin 23° 2'}{\sin 43° 20'} = \frac{182(0.3912)}{0.6862} = 103.8$$

Summarizing and rounding off the computed parts, we have

$B = 66° 20'$ $C = 70° 20'$ $c = 250$
$B' = 113° 40'$ $C' = 23° 00'$ $c' = 104$

Exercise 8-2

Draw each triangle approximately to scale. Then solve for the unknown parts without the use of logarithms. Express the results properly rounded off.

1. $a = 50, b = 26, A = 95°$
2. $a = 40, b = 22, A = 93°$
3. $b = 60, c = 82, B = 100°$
4. $b = 51, c = 51, C = 90°$
5. $a = 31, b = 26, B = 48°$
6. $a = 42, b = 53, A = 39°$
7. $b = 3.8, c = 2.7, C = 32°$
8. $b = 8.4, c = 9.8, C = 71°$
9. $a = 54, b = 61, A = 58°$
10. $b = 36, c = 48, B = 37°$

Use logarithms and solve each problem 11 through 20.

11. $b = 140, c = 203, C = 100° 10'$
12. $a = 602, c = 49.7, C = 52° 30'$
13. $a = 2434, b = 3462, B = 48° 32'$
14. $b = 0.6025, c = 0.8830, C = 32° 34'$

15. $b = 134.2$, $c = 158.6$, $B = 43° \ 7'$
16. $a = 56.15$, $b = 74.23$, $A = 50° \ 18'$
17. $b = 4798$, $c = 6540$, $C = 91° \ 31'$
18. $a = 0.8142$, $b = 0.6372$, $A = 71° \ 14'$
19. $a = 4.892$, $b = 7.540$, $A = 43° \ 19'$
20. $b = 298.4$, $c = 310.3$, $C = 68° \ 31'$

21. To determine the distance between two points A and B, separated by a lake, a surveyor measures a distance of 304 feet from A to C such that angle BAC is $40° \ 20'$. He then measures CB as 425 feet. Find the distance AB.

22. A pole leans southward and makes an angle of $60°$ with the horizontal. A prop 12 feet long supports the pole. If the base of the prop is 13 feet south of the pole, find the greater acute angle which the prop may make with the horizontal.

23. Town B is due east of town A and town C is in the direction N $68°$ E of town A. If towns A and C are 19 miles apart and towns B and C are 11 miles apart, find the distance between A and B.

24. A telephone pole is supported by two guy wires extending from the top of the pole to the ground on opposite sides of the pole. One wire is 50 feet long and makes an angle of $48°$ with the ground. The other wire is 46 feet long. Find the acute angle which the second wire makes with the ground. Find also the distance between the wires at ground level.

25. Do problem 24 if the guy wires are on the same side of the pole, the other data being unchanged.

8-3 The law of cosines

The law of sines is not directly applicable to cases III and IV. But the law of cosines, which we shall now derive, is applicable to these cases.

Theorem 8-2   *In any triangle the square of any side is equal to the sum of the squares of the other two sides minus twice the product of those sides and the cosine of the included angle. That is,*

$$a^2 = b^2 + c^2 - 2bc \cos A$$
$$b^2 = a^2 + c^2 - 2ac \cos B$$
$$c^2 = a^2 + b^2 - 2ab \cos C$$

Proof   Consider the triangle ABC (Fig. 8-14) where we have introduced a coordinate system with the origin at the vertex of angle A and the x axis along side c. Note that the coordinates of the vertices of the triangle are

$O(0,0)$ $B(c,0)$ $C(b \cos A, \ b \sin A)$

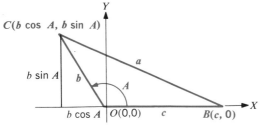

FIGURE 8-14

Although angle A is obtuse in the figure, these coordinates would be unchanged if A were an acute angle or a right angle. We now employ the distance formula (Sec. 1-5) to find the square of length a. Thus

$$
\begin{aligned}
a^2 &= (b \cos A - c)^2 + (b \sin A - 0)^2 \\
&= b^2 \cos^2 A - 2bc \cos A + c^2 + b^2 \sin^2 A \\
&= b^2 (\cos^2 A + \sin^2 A) - 2bc \cos A + c^2 \\
&= b^2 + c^2 - 2bc \cos A
\end{aligned}
$$

Clearly, we could place the origin of coordinates at each of the other vertices and obtain relations involving $\cos B$ and $\cos C$.

It is immediately evident that the third side of a triangle may be computed by the law of cosines if two sides and their included angle are known. The formulas show also that any angle may be found if three sides of the triangle are given. Hence the law of cosines is applicable to case III and to case IV. The cosine formulas, however, involving addition and subtraction, are not well suited to logarithmic computation. Because of this disadvantage, other laws, later to be derived, are usually preferable for handling cases III and IV. Still the law of cosines is of much importance, for it has numerous applications in situations where computations are not involved.

EXAMPLE 1   Two sides and the included angle of a triangle are measured to be 11 feet, 20 feet, and 112°. Find the length of the third side.

Solution   We let b, c, and A stand for the given parts as indicated in Fig. 8-15. Then using the law of cosines, we obtain

$$
\begin{aligned}
a^2 &= b^2 + c^2 - 2bc \cos A \\
&= (11)^2 + (20)^2 - 2(11)(20) \cos 112° \\
&= 121 + 400 - 440(-0.3746) \\
&= 521 + 164.8 = 685.8 \\
a &= 26 \quad \text{rounded off to two digits}
\end{aligned}
$$

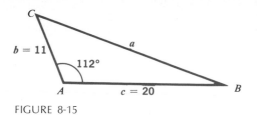

FIGURE 8-15

Notice that the reference angle of 112° is 68°, and therefore cos 112° = −cos 68° (Sec. 3-4). We remark also that the square root of a number may be found by the use of logarithms.

We can express the law of cosines in other useful forms by solving for the cosines of the angles. Thus,

$$\cos A = \frac{b^2 + c^2 - a^2}{2bc}$$

$$\cos B = \frac{a^2 + c^2 - b^2}{2ac}$$

$$\cos C = \frac{a^2 + b^2 - c^2}{2ab}$$

These formulas are convenient when the angles are to be found from the three known sides.

EXAMPLE 2   Given $a = 15.2$, $b = 20.3$, and $c = 30.0$, we find A, B, and C.

Solution   We shall find each of the angles by use of the law of cosines. We could, of course, employ the law of sines after one angle is found.

$$\cos A = \frac{(20.3)^2 + (30.0)^2 - (15.2)^2}{2(20.3)(30.0)} = \frac{1081}{1218} = 0.8875$$

$$A = 27° \ 30'$$

and

$$\cos B = \frac{(15.2)^2 + (30.0)^2 - (20.3)^2}{2(15.2)(30.0)} = \frac{719.8}{912.0} = 0.7883$$

$$B = 38° \ 00'$$

and

$$\cos C = \frac{(15.2)^2 + (20.3)^2 - (30.0)^2}{2(15.2)(20.3)} = \frac{-256.9}{617.1} = -0.4163$$

The cosine is negative; hence C is obtuse. We find the angle whose cosine is $+.4163$ and take its supplement to get C. Thus

$$C = 180° - 65° \, 20' = 114° \, 40'$$

Check

$$A + B + C = 27° \, 30' + 38° \, 00° + 114° \, 40' = 180° \, 10'$$

The check is satisfactory. The extra $10'$ in the sum of the angles is not surprising since the value of each angle is computed to the nearest integral multiple of $10'$.

Exercise 8-3

Find the unknown parts in each triangle in problems 1 through 15.

1. $a = 4.0$, $b = 5.0$, $C = 53°$
2. $b = 7.0$, $c = 10$, $A = 62°$
3. $A = 12$, $c = 17$, $B = 98°$
4. $a = 1.5$, $b = 2.4$, $C = 119°$
5. $b = 13$, $c = 24$, $A = 27°$
6. $a = 21$, $c = 32$, $B = 48°$
7. $b = 0.28$, $c = 0.43$, $A = 129°$
8. $b = 0.70$, $c = 0.40$, $A = 72°$
9. $a = 7.0$, $b = 8.0$, $c = 9.0$
10. $a = 34$, $b = 32$, $c = 41$
11. $a = 22$, $b = 26$, $c = 33$
12. $a = 21$, $b = 33$, $c = 26$
13. $a = 17.0$, $b = 24.0$, $c = 33.0$
14. $a = 3.00$, $b = 1.30$, $c = 1.60$
15. $a = 230$, $b = 306$, $c = 400$

16. Two sides of a parallelogram make an angle of $54°$. The lengths of the two sides are 21 and 33 feet. Find the length of each diagonal.
17. Two points A and B are on opposite sides of a pond. The distance from A to a third point C is measured to be 21.6 feet and the distance from C to B is measured to be 29.5 feet. Find the distance from A to B if angle ACB is $55° \, 30'$.
18. Two boats leave the same port at the same time. One travels 29.0 miles in the direction N $48° \, 20'$ E while the other travels 26.0 miles in the direction S $78° \, 10'$ E. What is the distance between the new positions of the boats?
19. Referring to Fig. 8-1, find two expressions for h^2 from the right triangles ABD and BCD. Equate the expressions and proceed to derive the law of cosines. Also make the derivation from Fig. 8-2, where A is an obtuse angle.

8-4   The area of a triangle

We shall now derive formulas for the area S of a triangle in terms of:

1. Two sides and the included angle
2. One side and three angles
3. Three sides

Theorem 8-3   *The area of a triangle is equal to half the product of two sides times the sine of the included angle. That is,*

$$S = \tfrac{1}{2}bc \sin A \qquad S = \tfrac{1}{2}ac \sin B \qquad S = \tfrac{1}{2}ab \sin C$$

Proof   The area of a triangle is equal to half the product of the base and the altitude. Hence for either triangle of Fig. 8-16 with h the altitude, we have

$$S = \tfrac{1}{2}hc = \tfrac{1}{2}bc \sin A$$

since $h = b \sin A$. The other two forms for the area evidently hold because any of the three angles could be labeled A.

Theorem 8-4   *The area of a triangle is equal to half the square of a side times the product of the sines of the adjacent angles divided by the sine of the opposite angle. That is,*

$$S = \frac{a^2 \sin B \sin C}{2 \sin A} = \frac{b^2 \sin C \sin A}{2 \sin B}$$

$$= \frac{c^2 \sin A \sin B}{2 \sin C}$$

Proof   If only one side and two (therefore three) angles are known, a second side is obtainable from the law of sines. That is,

$$\frac{b}{\sin B} = \frac{a}{\sin A} \qquad \text{and} \qquad b = \frac{a \sin B}{\sin A}$$

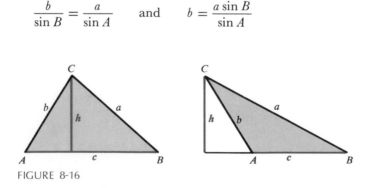

FIGURE 8-16

Then substituting for b in $S = \frac{1}{2}ab \sin C$ gives

$$S = \frac{a^2 \sin B \sin C}{2 \sin A}$$

The other two forms above for the area may be obtained by replacing a in this formula, in turn, by $\dfrac{b \sin A}{\sin B}$ and $\dfrac{c \sin A}{\sin C}$.

Theorem 8-5   *The area of a triangle in terms of its sides is given by*

$$S = \sqrt{s(s - a)(s - b)(s - c)}$$

where $s = \frac{1}{2}(a + b + c)$.

Proof   For this case we start with $S = \frac{1}{2}ab \sin C$. Then

$$
\begin{aligned}
S^2 &= \tfrac{1}{4}a^2b^2 \sin^2 C \\
&= \tfrac{1}{4}a^2b^2(1 - \cos^2 C) \\
&= \tfrac{1}{4}a^2b^2\left[1 - \left(\frac{a^2 + b^2 - c^2}{2ab}\right)^2\right] \qquad \text{by law of cosines} \\
&= \tfrac{1}{4}a^2b^2\left[\frac{4a^2b^2 - (a^2 + b^2 - c^2)^2}{4a^2b^2}\right] \\
&= \tfrac{1}{16}[2ab + (a^2 + b^2 - c^2)][2ab - (a^2 + b^2 - c^2)] \\
&= \tfrac{1}{16}[(a + b)^2 - c^2][c^2 - (a - b)^2] \\
&= \tfrac{1}{16}(a + b + c)(a + b - c)(c + a - b)(c - a + b)
\end{aligned}
$$

To simplify the right member, we set

$$a + b + c = 2s$$

Then subtracting $2a$, $2b$, and $2c$, in turn, from both sides gives

$$
\begin{aligned}
b + c - a &= 2(s - a) \\
a + c - b &= 2(s - b) \\
a + b - c &= 2(s - c)
\end{aligned}
$$

We may now write

$$S^2 = \frac{2s \cdot 2(s - c) \cdot 2(s - b) \cdot 2(s - a)}{16}$$

and hence

$$S = \sqrt{s(s-a)(s-b)(s-c)}$$

None of the formulas which we have derived gives the area of a triangle in terms of two sides and the angle opposite one of them. For these known parts the angles can first be found and then the area is obtainable through two sides and the included angle. This is the ambiguous case, however, and there is a possibility of two solutions.

EXAMPLE 1   If $a = 200$, $b = 300$, and $C = 40° \ 10'$, find the area of the triangle to three significant figures.

Solution   Using $s = \frac{1}{2}ab \sin C$, we have

$$S = \frac{1}{2}(200)(300) \sin 40° \ 10'$$
$$= \frac{1}{2}(200)(300)(0.6450)$$
$$= 19{,}400 \qquad \text{rounded off}$$

EXAMPLE 2   Find the area of the triangle with the sides $a = 20.55$, $b = 32.06$, and $c = 41.27$.

Solution   We use $S = \sqrt{s(s-a)(s-b)(s-c)}$.

$$
\begin{array}{ll}
a = 20.55 & s = 46.94 \\
b = 32.06 & s - a = 26.39 \\
c = 41.27 & s - b = 14.88 \\
\overline{2s = 93.88} & s - c = 5.67
\end{array}
$$

$$
\begin{array}{ll}
\log s = 1.6716 & \log S = 2.5096 \\
\log (s-a) = 1.4214 & \\
\log (s-b) = 1.1726 & S = 323.3 \\
\log (s-c) = 0.7536 & \\
\overline{\log S^2 = 5.0192} &
\end{array}
$$

Exercise 8-4

Find the area of each triangle. Use logarithms only where the computation will be facilitated.

1. $a = 12$, $b = 17$, $C = 150°$
2. $a = 12$, $b = 17$, $C = 30°$
3. $b = 50$, $c = 40$, $A = 48°$
4. $b = 3.2$, $c = 4.0$, $A = 115°$
5. $a = 60$, $B = 68°$, $C = 81°$
6. $b = 55$, $A = 80°$, $C = 98°$
7. $a = 41.3$, $B = 66° \ 10'$, $C = 78° \ 50'$

8. $b = 8.14$, $A = 28° 30'$, $C = 108° 20'$
9. $a = 7.0$, $b = 6.0$, $A = 64°$
10. $b = 14.4$, $c = 22.7$, $C = 102° 20'$
11. $a = 68.4$, $b = 79.5$, $A = 30° 40'$
12. $b = 30.8$, $c = 24.2$, $B = 97° 30'$
13. $a = 27$, $b = 36$, $c = 49$
14. $a = 9.00$, $b = 9.18$, $c = 7.34$
15. $a = 0.7521$, $b = 0.8731$, $c = 0.6222$

16. Denoting the length of each of the equal sides of isosceles triangle by a, and each of the equal angles by A, show that the area S is given by the formula

$$S = \tfrac{1}{2}a^2 \sin 2A$$

17. Verify each of the formulas for areas of triangles, for the case in which all sides are equal to a.

Chapter 8   Review exercise

Solve problems 1 through 8.

1. $b = 3.1$, $c = 3.2$, $A = 84°$. Find a.
2. $a = 64$, $b = 72$, $C = 91°$. Find c.
3. $a = 86.6$, $b = 92.8$, $A = 39° 0'$. Find c.
4. $a = 643$, $A = 36° 10'$, $C = 36° 30'$. Find c.
5. $a = 6150$, $c = 8457$, $A = 24° 41'$. Find C.
6. $c = 8054$, $A = 30° 16'$, $B = 43° 34'$. Find b.
7. $a = 21$, $b = 22$, $C = 99°$. Find c.
8. $a = 6.0$, $b = 8.3$, $c = 9.0$. Find C.

9. A railroad track crosses a highway at an angle of $73°$. A locomotive is 50 yards from the intersection when a car is 60 yards from the intersection. What then is the distance between the two vehicles?

10. A ship is anchored 25 miles S $20°$ E from a lighthouse. After moving 25 miles in the direction N $55°$ E, what is the ship's distance from the lighthouse?

11. The lengths of two sides of a parallelogram are 25 and 30 feet and the length of one of the two diagonals is 40 feet. Find the angles of the parallelogram.

12. Two men standing 90 yards apart on a straight level road sighted an airplane as it passed over the road between them. The angles of elevation of the airplane from the men when sighted were $40°$ and $35°$, respectively. Find the height of the airplane above the road.

13. A triangle has $a = 12$, $b = 16$, $A = 35°$. Find the other parts of the triangle.

14. Two adjacent sides of a parallelogram are 15 feet and 24 feet and the angle between them is 51°. Find the length of each diagonal of the parallelogram.

15. The length of one diagonal of a parallelogram is 33 feet and it makes angles of 26° and 30° with the sides. Find the lengths of the sides.

vectors and applications

9-1 Vectors

Many physical quantities possess the properties of magnitude and direction. A quantity of this kind is called a *vector quantity*. A force, for example, is characterized by its magnitude and direction of action. The force would not be completely specified by one of these properties without the other. The velocity of a moving body is determined by its speed (magnitude) and direction of motion. Acceleration and displacement are other examples of vector quantities.

To obtain a geometric representation of a vector quantity, we employ a directed line segment (Sec. 1-2) whose length and direction represent the magnitude and direction, respectively, of the vector quantity. In order to get a working basis for investigating problems involving vector quantities, we shall introduce certain definitions and establish a number of useful theorems.

Definition 9-1   *A directed line segment, when used to represent a vector quantity, is called a* vector.

We shall denote a vector by a boldface letter or by giving its starting point and its ending point. Thus the vector (arrow) in Fig. 9-1 is drawn from O to P and we let **OP** or **A** indicate the vector. The point O is called the *foot* of the vector and the point P is called the *head*. The vectors **B** and $-\mathbf{B}$ in the figure have the same length as **A.** The vectors **A** and **B** have the same direction but $-\mathbf{B}$ is oppositely directed. The three vectors are related in accordance with the following definition.

Definition 9-2   *Two vectors* **A** *and* **B** *are* equal $(\mathbf{A} = \mathbf{B})$ *if they have the same length and direction. The* negative *of a vector* **B,** *denoted by* $-\mathbf{B}$, *is a vector having the same length as* **B** *but pointing in the opposite direction.*

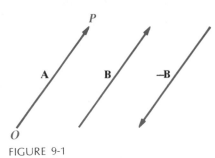

FIGURE 9-1

In view of this definition, we note that vectors of the same length and different directions are not equal, and vectors of the same direction and different lengths are not equal.

9-2   Operations on vectors

The operations of addition, subtraction, and multiplication of vectors are defined differently from the corresponding operations on real numbers. The new operations are designed so as to establish an appropriate theory for studying forces, velocities, and other physical concepts.

Definition 9-3   *Let* **A** *and* **B** *denote vectors and draw from the head of* **A** *a vector equal to* **B**. *Then the sum* (**A** + **B**) *of* **A** *and* **B** *is the vector extending from the foot of* **A** *to the head of* **B**.

This definition is illustrated in Fig. 9-2. The sum of the vectors is called the *resultant* and each of the vectors forming the sum is called a *component*. The triangle formed by the three vectors **A, B,** and **A** + **B** is called a *vector triangle*. This method of adding vectors is used in physics, where it is shown, for example, that two forces applied at a point of a body have exactly the same effect as a single force equal to their resultant.

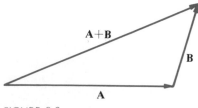

FIGURE 9-2

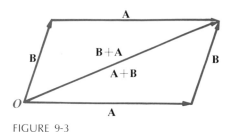

FIGURE 9-3

Theorem 9-1   *Vector addition is commutative.*

Proof   We need to show that if **A** and **B** are any two vectors, then **A** + **B** = **B** + **A.** For this purpose we draw **A** and **B** from the point O (Fig. 9-3), and then complete a parallelogram with the vectors forming adjacent sides. Since the opposite sides of a parallelogram are equal and parallel, we see that the foot of **B** in the lower part of the figure is at the head of **A.** Hence the sum **A** + **B** is along the diagonal extending from O. And from the upper part of the figure, we see that **B** + **A** is along the same diagonal. We conclude then that **A** + **B** = **B** + **A,** which means the vectors are commutative with respect to addition.

Theorem 9-2   *Vectors obey the associative law of addition.*

Proof   Given three vectors **A, B,** and **C,** we are to prove that

$$(\mathbf{A} + \mathbf{B}) + \mathbf{C} = \mathbf{A} + (\mathbf{B} + \mathbf{C})$$

From Fig. 9-4 we observe that **OE** = **A** + **B** and that **C** added to this sum yields

$$\mathbf{OF} = (\mathbf{A} + \mathbf{B}) + \mathbf{C}$$

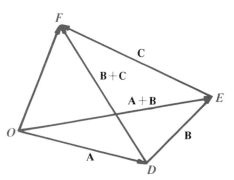

FIGURE 9-4

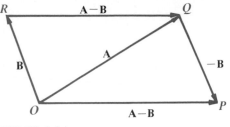

FIGURE 9-5

Similarly, $\mathbf{DF} = \mathbf{B} + \mathbf{C}$ and adding this sum to $\mathbf{A}$ gives

$$\mathbf{OF} = \mathbf{A} + (\mathbf{B} + \mathbf{C})$$

Hence the sums $(\mathbf{A} + \mathbf{B}) + \mathbf{C}$ and $\mathbf{A} + (\mathbf{B} + \mathbf{C})$ have the same resultant, and consequently vector addition is associative.

Definition 9-4   *A vector **B** subtracted from a vector **A** is equal to the sum of **A** and the negative of **B**. That is,*

$$\mathbf{A} - \mathbf{B} = \mathbf{A} + (-\mathbf{B})$$

Referring to the parallelogram *OPQR* in Fig. 9-5, we observe that the vector from *O* to *P* is equal to $\mathbf{A} - \mathbf{B}$, and the vector from *R* to *Q* is also equal to $\mathbf{A} - \mathbf{B}$. So, alternatively, if $\mathbf{A}$ and $\mathbf{B}$ are drawn from a common point, the vector from the head of $\mathbf{B}$ to the head of $\mathbf{A}$ is equal to $\mathbf{A} - \mathbf{B}$.

When numbers and vectors are involved in a problem or discussion, the numbers are sometimes called *scalars* to distinguish them from vectors.

Definition 9-5   *The product of a scalar m and a vector **A**, expressed by m**A**, is a vector m times as long as **A**, and has the direction of **A** if m is positive and the opposite direction if m is negative.*

This definition is illustrated in Fig. 9-6, where the scalar m has the values -2 and $\frac{3}{4}$.

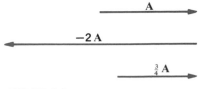

FIGURE 9-6

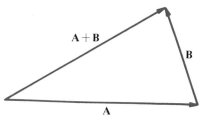

FIGURE 9-7

Theorem 9-3   *If m and n are scalars and* **A** *and* **B** *are vectors, then*

$$(m + n)\mathbf{A} = m\mathbf{A} + n\mathbf{A} \tag{1}$$

and

$$m(\mathbf{A} + \mathbf{B}) = m\mathbf{A} + m\mathbf{B} \tag{2}$$

Proof   We leave the proof of equation (1) to the student. To establish equation (2), we note that **A, B,** and **A + B** form the sides of a triangle (Fig. 9-7). If each of these vectors is multiplied by a nonzero scalar m, the products $m\mathbf{A}$, $m\mathbf{B}$, and $m(\mathbf{A} + \mathbf{B})$ can be placed so as to form a triangle in which $m(\mathbf{A} + \mathbf{B}) = m\mathbf{A} + m\mathbf{B}$ (Fig. 9-8). Hence, as expressed by equations (1) and (2), scalars and vectors obey the distributive law of multiplication.

Let us seek an interpretation of the difference of a vector and itself and the product of zero and a vector. That is, if **A** is a vector, what meaning should be given to $\mathbf{A} - \mathbf{A}$ and $(0)\mathbf{A}$? For these quantities to be vectors, Definitions 9-4 and 9-5 require the length in each case to be equal to zero. To handle a situation of this kind, it is customary to enlarge the concept of a vector to include one of zero length, which is called the *zero vector*.

It is sometimes desirable to find two vectors whose sum is equal to a given vector. The given vector is then said to be *resolved* into two components. The components may be along any two directions in a plane containing the

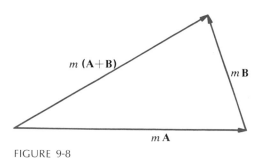

FIGURE 9-8

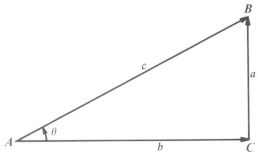

FIGURE 9-9

given vector. A graphical construction of the components may be obtained by forming a vector triangle of which the given vector is a side. The components are then along the other sides.

We shall show how to resolve a vector **AB** (Fig. 9-9) into a pair of perpendicular components. Let θ be the angle which **AB** makes with a chosen direction. Then complete a right triangle ABC by drawing a line from B perpendicular to the chosen direction. Clearly the vector **AB** is equal to the sum of the vectors **AC** and **CB**. Further, letting the magnitudes of **AC**, **CB**, and **AB** be b, a, and c, respectively, we may write

$$a = c \sin \theta \quad \text{and} \quad b = c \cos \theta$$

9-3 Applications of vectors

In this section we shall deal with vector problems whose solutions involve right triangles. The illustrative examples should be followed closely.

EXAMPLE 1   A force of 180 pounds acts perpendicularly to a force of 270 pounds. Find the magnitude of the resultant and the angle between the resultant and the direction of the 270-pound force.

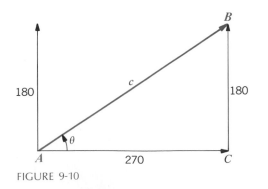

FIGURE 9-10

Solution   Let **AC** and **CB** represent the forces, **AB** the resultant, θ the angle between **AC** and **AB,** and c the magnitude of **AB** (Fig. 9-10). Then

$$\tan \theta = \tfrac{180}{270} = 0.6667$$
$$\theta = 33° \ 40'$$
$$c = 270 \sec 33° \ 40'$$
$$= 270(1.20) = 324$$

These results, rounded off, are $\theta = 34°$ and $c = 320$ pounds.

EXAMPLE 2   A body weighing 320 pounds is on a smooth (friction negligible) inclined plane which makes an angle of 28° with the horizontal. What force parallel to the plane will just keep the body from slipping down the plane? With what force does the body push against the plane?

Solution   The 320-pound weight acts vertically downward and is represented by the vector **AB** (Fig. 9-11). The components of **AB** perpendicular and parallel to the plane are, respectively, **AC** and **CB**. The component **AC** represents the push of the body against the plane and the component **CB** represents the force tending to slide the body down the plane. Noting that the angle between **AB** and **AC** is 28° and indicating the magnitude of **CB** by a and the magnitude of **AC** by b, we have

$$a = 320 \sin 28° \qquad\qquad b = 320 \cos 28°$$
$$= 320(0.470) \qquad\qquad = 320(0.883)$$
$$= 150 \quad \text{rounded off} \qquad = 280 \quad \text{rounded off}$$

From these results, we find the components of the weight of the body to be 150 pounds parallel to the plane and 280 pounds perpendicular to the plane. Hence a force of 150 pounds upward and parallel to the plane will just keep the body from slipping, and the body pushes against the plane with a force of 280 pounds.

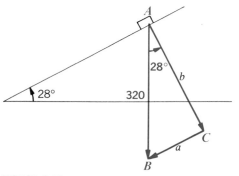

FIGURE 9-11

EXAMPLE 3   A body weighing 400 pounds is held in equilibrium (at rest) by two ropes, one making an angle of 25° with the vertical and the other an angle of 40° with the vertical. Find the force exerted on the body by each rope.

Solution   To find the unknown forces, we apply the following conditions.

1.   If a body is in equilibrium under a number of forces, the sum of the forces must be equal to zero; otherwise the body would move in the direction of the sum of the forces.
2.   If a body is in equilibrium under a number of forces, the sum of the components of the forces in any chosen direction must be equal to zero.

We let f_1 and f_2 stand for the magnitudes of the forces exerted by the ropes (Fig. 9-12). Then we note that the vertical components of the forces are both upward and that their horizontal components are in opposite directions. The pull of gravity on the body is, of course, vertically downward. So, choosing the positive direction upward and equating the vertical and horizontal components of the three forces involved to zero, we obtain the system of equations

$$f_1 \cos 25° + f_2 \cos 40° - 400 = 0 \tag{3}$$
$$f_1 \sin 25° - f_2 \sin 40° = 0 \tag{4}$$

Multiplying the members of equation (3) by $\sin 40°$ and equation (4) by $\cos 40°$, we pass to the equivalent system

$$f_1 \cos 25° \sin 40° + f_2 \cos 40° \sin 40° = 400 \sin 40° \tag{5}$$
$$f_1 \sin 25° \cos 40° - f_2 \sin 40° \cos 40° = 0 \tag{6}$$

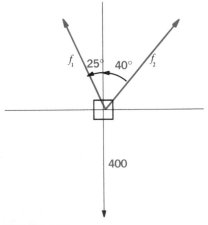

FIGURE 9-12

Adding corresponding members of equations (5) and (6), we get

$$f_1(\cos 25° \sin 40° + \sin 25° \cos 40°) = 400 \sin 40°$$

Thus

$$f_1 \sin 65° = 400 \sin 40°$$

and

$$f_1 = 284 \text{ pounds}$$

Then this result and equation (4) yield

$$f_2 = \frac{284 \sin 25°}{\sin 40°}$$

$$= 187 \text{ pounds}$$

Rounded off to two significant figures, the final results are $f_1 = 280$ pounds and $f_2 = 190$ pounds.

Vectors also have applications in aviation. We let **OB** (Fig. 9-13) represent the speed of an airplane relative to the air and the direction in which the airplane points, or heads. If the air were not in motion, the speed and heading would bring the airplane from O to B in one unit of time. If, however, **BC** represents the speed and direction of the wind, the airplane would not travel from O to B, but along the line from O to C, reaching C in one unit of time. Hence the vector **OC,** the sum of the vectors **OB** and **BC,** represents the path and speed of the airplane relative to the ground. The magnitude of **OC** is called the *groundspeed,* and the magnitude of **OB** is called the *airspeed.*

In aviation the bearing of a point, or direction of a course of flight, is measured in a clockwise direction from the north. Thus in Fig. 9-14, the direction of OA is 55°, that of OB is 220°, and that of OC is 315°.

EXAMPLE 4   A pilot whose plane has an airspeed of 180 miles per hour wishes to travel in a due-east direction. If the wind is blowing north at 44 miles

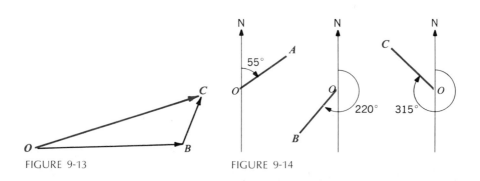

FIGURE 9-13   FIGURE 9-14

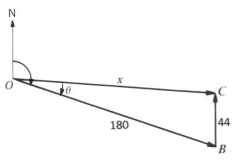

FIGURE 9-15

per hour, find the direction in which the pilot should head the plane and the resulting groundspeed of the plane.

Solution   We let the vector **OB** (Fig. 9-15) represent the airspeed and direction of heading, **BC** the speed and direction of the wind, **OC** the groundspeed and path of travel, and θ the angle BOC. Then we have

$$\sin \theta = \tfrac{44}{180} = 0.2444$$
$$\theta = 14° \ 10'$$

Letting x stand for the length of **OC**, we get

$$x = 44 \cot 14° \ 10'$$
$$= 44(3.96) = 174.24$$

Rounding off, we find the heading to be $104°$ and the groundspeed 170 miles per hour.

Exercise 9-1

Assume that all numbers in the following problems are of two-figure accuracy and that each angle is given to the nearest degree.

1. A force of 160 pounds and another of 200 pounds act at right angles to each other. Find the magnitude of the resultant and angle which it makes with each force.
2. A force of 210 pounds acts in a due-east direction and another force of 330 pounds acts due south. Find the direction and magnitude of the resultant.
3. A boat is going south at the rate of 20 miles per hour. A man walks east across the deck at 3.1 miles per hour. Find the magnitude and direction of the man's velocity relative to the surface of the water.
4. Find the horizontal and vertical components of a 320-pound force which makes an angle of 32° with the vertical.

5. A bullet is fired from a gun with a muzzle velocity of 2100 feet per second. The gun makes an angle of 36° with the horizontal. Find the magnitudes of the horizontal and vertical components of the velocity of the bullet as it leaves the gun.

6. Suppose a 360-pound block is on a smooth plane inclined at 30° to the horizontal. What force parallel to the plane will just keep the block from sliding down the plane? With what force then does the block push against the plane?

7. A car, standing on a pavement which makes an angle of 15° with the horizontal, is kept at rest by a force of 430 pounds parallel to the pavement. Find the weight of the car and the force it exerts against the pavement.

8. If a force of 190 pounds parallel to a smooth inclined plane is just sufficient to keep a 350-pound weight from sliding down the plane, find the angle of inclination of the plane.

9. A man slides a 460-pound weight up a smooth inclined plane. If he can push with a force of 160 pounds, find the greatest angle which the plane may make with the horizontal.

10. Suppose a car is on a portion of pavement which makes an angle of 9° with the horizontal. If a horizontal force of 200 pounds, in the vertical plane through the pavement, will move the car slowly, find the weight of the car.

11. An airplane is headed due east with an airspeed of 260 miles per hour. A wind is blowing due north at 50 miles per hour. Find the groundspeed and direction which the plane takes.

12. The airspeed of an airplane is 240 miles per hour. A wind is blowing from the west at 52 miles per hour. In what direction should the pilot head in order to travel due north?

13. A pilot, flying in a wind which is blowing 38 miles per hour due south, finds he moves due east when the plane is headed in the direction 67°. Find the airspeed and groundspeed of the plane.

14. An airplane is headed in the direction 74° with an airspeed of 180 miles per hour. A wind blowing due south causes the airplane to travel due east. Find the groundspeed of the airplane and the speed of the wind.

15. Two children are carrying an object. One exerts a force of 42 pounds at an angle of 42° with the vertical and the other a force of 62 pounds at an angle of 27° with the vertical. Find the weight of the object.

16. A crate is supported by two ropes fastened to it. One rope makes an angle of 21° with the vertical and has a tension of 270 pounds and the other rope is in a horizontal position. Find the weight of the crate and the tension in the horizontal rope.

17. Two men are carrying a box. One man exerts a force of 170 pounds at an angle of 22° with the vertical and the other man exerts a force at an angle of 33° with the vertical. Find the force which the other man exerts and the weight of the box.

18. A body weighing 1000 pounds is held in equilibrium by two ropes, one making an angle of 28° with the vertical and the other making an angle of 36° with the vertical. Find the tension in each rope.

19. The buoyancy of a certain balloon in air is 340 pounds. The balloon is kept from rising by two ropes which make angles with the vertical of 27° and 40°. Find the tension in each rope.

20. An airplane is flying at a rate of 440 miles per hour in the direction 38°. Find the components of its velocity in the directions 85° and 355°.

9-4   Vector problems involving oblique triangles

The problems of the previous section involved right triangles. We now pass to problems which lead to oblique vector triangles.

EXAMPLE 1   A pilot heads his plane in the direction 60°, but a wind blowing in the direction 210° causes his course to be in the direction 70°. If he travels on this course 160 miles in 1 hour, find his airspeed and the speed of the wind.

Solution   In the vector triangle ABC (Fig. 9-16),

AB represents the groundspeed and direction of flight.
CB represents the speed and direction of the wind.
AC represents the airspeed and direction of heading.

From the given directions, we find the angles of triangle ABC to be $A = 10°$, $C = 30°$, and $B = 140°$. Hence, by the law of sines,

$$\frac{a}{\sin 10°} = \frac{160}{\sin 30°} \quad \text{and} \quad \frac{b}{\sin 140°} = \frac{160}{\sin 30°}$$

These equations yield $a = 56$ and $b = 210$, and therefore the speed of the wind is 56 miles per hour and the airspeed is 210 miles per hour.

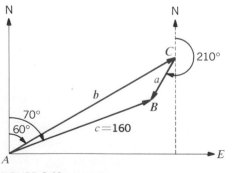

FIGURE 9-16

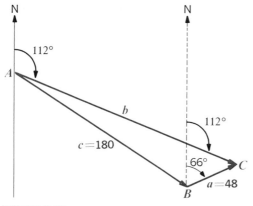

FIGURE 9-17

EXAMPLE 2   A plane with an airspeed of 180 miles per hour is flying in a 48-mile-per-hour wind blowing in the direction 66°. The plane is headed so that its actual course is in the direction 112°. Find the heading and groundspeed of the plane.

Solution   In the vector triangle ABC (Fig. 9-17),

AB represents the heading and airspeed.
BC represents the direction and speed of the wind.
AC represents the direction of flight and groundspeed.

Noting that angle C of triangle ABC is equal to 46°, we apply the law of sines and get

$$\frac{\sin A}{48} = \frac{\sin 46°}{180} \quad \text{and} \quad \frac{b}{\sin B} = \frac{48}{\sin A}$$

From these equations we find $A = 11°$ and $b = 210$. Hence the heading of the plane is 123° and groundspeed is 210 miles per hour.

EXAMPLE 3   Three forces of 360, 400, and 580 pounds acting at a point are in equilibrium. Find to the nearest degree the angles between the directions of the forces.

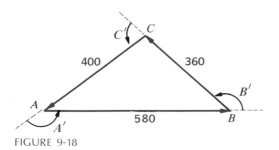

FIGURE 9-18

Solution   The vectors representing the forces may be placed so as to form the triangle ABC (Fig. 9-18). The interior angles of the triangle, by the law of cosines, are $A = 38°$, $B = 43°$, and $C = 99°$. The exterior angles, which give the angles between the directions of the forces, are therefore $A' = 142°$, $B' = 137°$, and $C' = 81°$.

Exercise 9-2

Assume that all numbers in the following problems are of two-figure accuracy and that each is expressed to the nearest degree.

1. A force of 170 pounds makes an angle of 78° with a second force. The resultant of the two forces makes an angle of 32° with the second force. Find the magnitudes of the second force and the resultant.

2. A boat, headed in the direction N 36° E, moves in the direction N 55° E in crossing a river. If the rate of the boat in still water is 4.0 miles per hour and if the river flows due south, find the speed of the current and the actual speed of the boat in crossing.

3. A pilot, flying with an airspeed of 280 miles per hour, heads his plane in the direction 150°. If the wind is blowing due east and the flight course is in the direction 140°, find the windspeed and the groundspeed of the plane.

4. A pilot heads his plane in the direction 80°. A 45-mile-per-hour wind blowing in the direction 45° causes the plane to move in the direction 75°. Find the airspeed and the groundspeed of the plane.

5. In crossing a river a boat heads in the direction N 30° E and has a groundspeed of 5.6 miles per hour. The rate of the boat in still water is 8.0 miles per hour, and the river flows due south. Find the bearing of the path of the boat and the smaller of the two possible rates of flow.

6. A pilot wishes to travel in the direction 90°. His airspeed is 150 miles per hour, and a 35-mile-per-hour wind is blowing in the direction 38°. Find the direction in which he should head his plane, and find his speed relative to the ground.

7. A pilot flying with an airspeed of 560 miles per hour travels in the direction 140°. If the wind is blowing due east at 56 miles per hour, find the heading and groundspeed of the plane.

8. A force of 50 pounds makes an angle of 67° with a force of 70 pounds. Find the magnitude of the resultant of the two forces.

9. A plane with an airspeed of 170 miles per hour heads in the direction 67°. It is carried from this course by a 42-mile-per-hour wind blowing in the direction 0°. Find the direction in which the plane travels and the groundspeed of the plane.

10. A force of 52 pounds just balances two forces of 31 pounds and 42 pounds. Find the angles between the directions of the forces.

11. An airplane heads due south with an airspeed of 210 miles per hour. The wind is blowing in the direction 122° at 40 miles per hour. Find the distance and direction of the plane from its starting point at the end of 2 hours.

12. A pilot wishes to fly a distance of 180 miles in a due-south direction. The wind is blowing in the direction 60° at 36 miles per hour. Find the necessary heading and airspeed for the pilot to reach the destination in 1 hour.

13. An airplane with an airspeed of 240 miles per hour is flying in a 36-mile-per-hour wind blowing in the direction 76°. The plane is headed so that the actual course is in the direction 123°. Find the heading and groundspeed of the plane.

14. A river 100 feet wide is flowing due east at 3.0 feet per second. A boat with a rowing rate of 4.0 feet per second in still water goes directly from a point on the south bank to a point on the opposite side 200 feet downstream. Find the heading of the boat and its groundspeed.

9-5   Vectors in a rectangular coordinate plane*

In our consideration of vectors thus far we have not used a coordinate system. Many operations on vectors, however, can be carried out advantageously by the aid of a coordinate system. To begin our study of vectors in a coordinate plane, we introduce two special vectors each of unit length. One of the vectors, denoted by **i,** has the direction of the positive x axis; the other vector, denoted by **j,** has the direction of the positive y axis. Each of these vectors, as well as any vector of unit length, is said to be a *unit vector.*

Since vectors of the same length and same direction are equal (Definition 9-2), each of the vectors **i** and **j** may extend from any chosen point of the coordinate plane. But it is usually convenient to place them so that they extend from the origin (Fig. 9-19).

The product m**i** is a vector of length m units and has the direction of **i** if m is positive and the opposite direction if m is negative (Definition 9-5). A similar statement applies to the product m**j**. Using these facts, we shall point out that any vector can be expressed in terms of the unit vectors **i** and **j**. Let **V** be a vector from the origin to the point (a,b), as shown in Fig. 9-20. Clearly a**i** and b**j** are the horizontal and vertical components of the given vector, and therefore $\mathbf{V} = a\mathbf{i} + b\mathbf{j}$. The vector a**i** is called the x *component* of **V** and b**j** the y *component.*

Continuing with the vector **V**, we make the following observations. Since the length of the x component of **V** is a and the length of the y component

*Parts of the remainder of this chapter were taken from Gordon Fuller, "Analytic Geometry," Addison-Wesley, Reading, Mass., 1967. Permission to use this material, without charge, was granted by the Addison-Wesley Publishing Company, Inc.

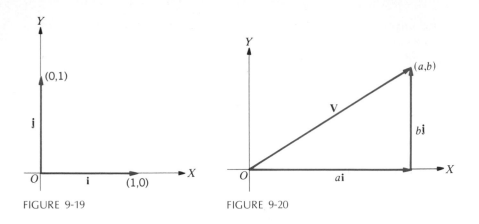

FIGURE 9-19   FIGURE 9-20

is b, we may employ the Pythagorean theorem to find the length or magnitude of **V**. Thus, denoting the length by $|\mathbf{V}|$, we have

$$|\mathbf{V}| = \sqrt{a^2 + b^2}$$

If **V** is divided by $|\mathbf{V}|$, the result is a unit vector with the same direction as that of **V**. The length of $\mathbf{V} = 3\mathbf{i} - 4\mathbf{j}$, for example, is

$$|\mathbf{V}| = \sqrt{9 + 16} = 5$$

and

$$\frac{\mathbf{V}}{|\mathbf{V}|} = \frac{3\mathbf{i} - 4\mathbf{j}}{5} = \frac{3}{5}\mathbf{i} - \frac{4}{5}\mathbf{j}$$

is a unit vector having the same direction as $3\mathbf{i} - 4\mathbf{j}$.

Theorem 9-4   *If the vectors $\mathbf{V}_1$ and $\mathbf{V}_2$, in terms of their x components and y components, are*

$$\mathbf{V}_1 = a_1\mathbf{i} + b_1\mathbf{j} \quad \text{and} \quad \mathbf{V}_2 = a_2\mathbf{i} + b_2\mathbf{j}$$

then

$$\mathbf{V}_1 + \mathbf{V}_2 = (a_1 + a_2)\mathbf{i} + (b_1 + b_2)\mathbf{j} \tag{7}$$

and

$$\mathbf{V}_1 - \mathbf{V}_2 = (a_1 - a_2)\mathbf{i} + (b_1 - b_2)\mathbf{j} \tag{8}$$

Proof   If the foot of $\mathbf{V}_1$ is at the origin, the head will be at the point (a,b). Then if the foot of $\mathbf{V}_2$ is at the head of $\mathbf{V}_1$, the head will be at the point $(a_1 + a_2, b_1 + b_2)$. And the vector from the origin to this point is expressed by $(a_1 + a_2)\mathbf{i} + (b_1 + b_2)\mathbf{j}$, which, by Definition 9-3, is equal to $\mathbf{V}_1 + \mathbf{V}_2$.

Although formula (7) follows readily from the definition of the sum of two vectors, it is worth noting that the formula can be established by the use of the commutative and associative properties for the addition of vectors (Theorems 9-1 and 9-2). Thus

$$\mathbf{V}_1 + \mathbf{V}_2 = \mathbf{V}_1 + (a_2\mathbf{i} + b_2\mathbf{j})$$
$$= (\mathbf{V}_1 + a_2\mathbf{i}) + b_2\mathbf{j}$$
$$= [(a_1\mathbf{i} + b_1\mathbf{j}) + a_2\mathbf{i}] + b_2\mathbf{j}$$
$$= (a_1\mathbf{i} + a_2\mathbf{i}) + (b_1\mathbf{j} + b_2\mathbf{j})$$
$$= (a_1 + a_2)\mathbf{i} + (b_1 + b_2)\mathbf{j}$$

We leave the proof of formula (8) to the student.

EXAMPLE 1   Vectors are drawn from the origin to the points $P(3,-2)$ and $Q(1,5)$. Indicating these vectors by $\mathbf{OP} = \mathbf{A}$ and $\mathbf{OQ} = \mathbf{B}$, find $\mathbf{A} + \mathbf{B}$ and $\mathbf{A} - \mathbf{B}$.

Solution   The given vectors, in terms of their x components and y components, are

$$\mathbf{A} = 3\mathbf{i} - 2\mathbf{j} \quad \text{and} \quad \mathbf{B} = \mathbf{i} + 5\mathbf{j}$$

Then, by the preceding theorem,

$$\mathbf{A} + \mathbf{B} = (3 + 1)\mathbf{i} + (-2 + 5)\mathbf{j}$$
$$= 4\mathbf{i} + 3\mathbf{j}$$

and

$$\mathbf{A} - \mathbf{B} = (3 - 1)\mathbf{i} + (-2 - 5)\mathbf{j}$$
$$= 2\mathbf{i} - 7\mathbf{j}$$

These two results are pictured in Fig. 9-21. The vector $\mathbf{A} - \mathbf{B}$, as we pointed out by use of Fig. 9-5, is equal to the vector extending from the

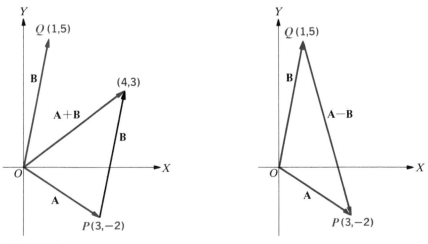

FIGURE 9-21

head of **B** to the head of **A**. The foot of **A** − **B**, in the figure, is not at the origin. But a vector equal to **A** − **B** with its foot at the origin would have $(2,-7)$ as the coordinates of its head.

EXAMPLE 2   Find the coordinates of the midpoint of the line segment joining the points $P(-2,4)$ and $Q(8,2)$.

Solution   We first find the vector from the origin to the midpoint of the line segment. This vector is equal to the vector from the origin to P plus half the vector from P to Q (Fig. 9-22). Indicating the vectors from the origin to P and Q by **A** and **B**, respectively, we have

$$\mathbf{A} = -2\mathbf{i} + 4\mathbf{j}$$
$$\mathbf{B} = 8\mathbf{i} + 2\mathbf{j}$$
$$\mathbf{B} - \mathbf{A} = 10\mathbf{i} - 2\mathbf{j}$$

The desired vector **V** then is

$$\mathbf{V} = \mathbf{A} + \tfrac{1}{2}(\mathbf{B} - \mathbf{A})$$
$$= (-2\mathbf{i} + 4\mathbf{j}) + \tfrac{1}{2}(10\mathbf{i} - 2\mathbf{j})$$
$$= 3\mathbf{i} + 3\mathbf{j}$$

This result shows that the head of **V** is at the point $(3,3)$, and these are the coordinates of the midpoint of PQ.

EXAMPLE 3   Find the vectors from the origin to the trisection points of the line segment joining the points $P(1,3)$ and $Q(4,-3)$. Give the coordinates of the trisection points.

Solution   One of the required vectors is equal to the vector from the origin to P plus one-third of the vector from P to Q (Fig. 9-23). The other required vector is equal to the vector from the origin to P plus two-thirds

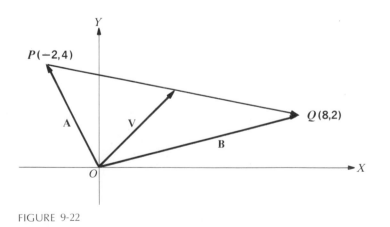

FIGURE 9-22

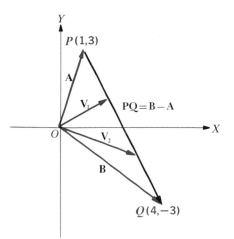

FIGURE 9-23

of the vector from P to Q. Denoting the vectors from the origin to P and Q by $\mathbf{A}$ and $\mathbf{B}$, respectively, we write

$$\mathbf{A} = \mathbf{i} + 3\mathbf{j}$$
$$\mathbf{B} = 4\mathbf{i} - 3\mathbf{j}$$
$$\mathbf{B} - \mathbf{A} = 3\mathbf{i} - 6\mathbf{j}$$

Hence one of the required vectors $\mathbf{V}_1$ is

$$\mathbf{V}_1 = \mathbf{A} + \tfrac{1}{3}(\mathbf{B} - \mathbf{A})$$
$$= (\mathbf{i} + 3\mathbf{j}) + \tfrac{1}{3}(3\mathbf{i} - 6\mathbf{j})$$
$$= 2\mathbf{i} + \mathbf{j}$$

The other vector $\mathbf{V}_2$ is

$$\mathbf{V}_2 = (\mathbf{i} + 3\mathbf{j}) + \tfrac{2}{3}(3\mathbf{i} - 6\mathbf{j})$$
$$= 3\mathbf{i} - \mathbf{j}$$

The vectors $2\mathbf{i} + \mathbf{j}$ and $3\mathbf{i} - \mathbf{j}$ tell us that the coordinates of the trisection points are $(2,1)$ and $(3,-1)$.

Exercise 9-3

In each problem 1 through 4, let vectors extend from the origin to the points with the given coordinates. Then find the sum of the vectors. Also subtract the second vector from the first. Draw all vectors.

1. $P(2,3)$, $Q(-4,5)$ **2.** $P(5,0)$, $Q(0,4)$
3. $P(3,-2)$, $Q(-1,-4)$ **4.** $P(6,7)$, $Q(-5,-5)$

Determine a unit vector having the direction of the vector in each problem 5 through 10.

5. $3\mathbf{i} + 4\mathbf{j}$ **6.** $12\mathbf{i} - 5\mathbf{j}$ **7.** $3\mathbf{i} - 12\mathbf{j}$
8. $\mathbf{i} + 2\mathbf{j}$ **9.** $2\mathbf{i} - 3\mathbf{j}$ **10.** $8\mathbf{i} + 15\mathbf{j}$

Use the method of Example 2 above to find the coordinates of the midpoint of the line segment joining P and Q in each problem 11 through 14.

11. $P(3,6)$, $Q(5,-8)$ **12.** $P(4,7)$, $Q(-2,-3)$
13. $P(-3,2)$, $Q(6,-4)$ **14.** $P(-2,-3)$, $Q(5,4)$

15. Use vectors, as in the preceding problems, to show that the coordinates of the midpoint of the line segment joining $P(a_1,b_1)$ and $Q(a_2,b_2)$ are

$$\left(\frac{a_1 + a_2}{2}, \frac{b_1 + b_2}{2}\right)$$

That is, the abscissa of the midpoint of a line segment is half the sum of the abscissas of the end points; the ordinate is half the sum of the ordinates of the end points.

Find the coordinates of the trisection points of the line segment joining P and Q in each problem 16 through 19. Use the method of illustrative Example 3 above.

16. $P(-3,-3)$, $Q(6,3)$ **17.** $P(-6,3)$, $Q(3,6)$
18. $P(-1,4)$, $Q(5,-2)$ **19.** $P(-3,-5)$, $Q(2,7)$

20. By the use of vectors show that the coordinates of the trisection points of the line segment joining $P(a_1,b_1)$ and $Q(a_2,b_2)$ are

$$\left(\frac{2a_1 + a_2}{3}, \frac{2b_1 + b_2}{3}\right) \quad \text{and} \quad \left(\frac{a_1 + 2a_2}{3}, \frac{b_1 + 2b_2}{3}\right)$$

9-6   The scalar product of two vectors

So far we have not defined a product of two vectors. Actually two kinds of vector products are important in physics, engineering, and other fields. We shall introduce the simpler of the two products.

Definition 9-6   *The* scalar product *of two vectors* **A** *and* **B**, *denoted by* **A** · **B**, *is the product of their lengths times the cosine of the angle* θ *between them. That is,*

$$\mathbf{A} \cdot \mathbf{B} = |\mathbf{A}|\,|\mathbf{B}|\cos\theta$$

The name scalar is used because the product is a scalar quantity. The product is also called the *dot product*. It makes no difference whether the angle θ is positive or negative, since $\cos \theta = \cos(-\theta)$. However, we shall restrict θ to the interval $0° \leq \theta \leq 180°$. The angle is equal to $0°$ if A and B point in the same direction, and it is equal to $180°$ if they point oppositely.

Since $\cos 90° = 0$ and $\cos 0° = 1$, it is evident that the scalar product of two perpendicular vectors is zero, and the scalar product of two vectors in the same direction is the product of their lengths. The dot product of a vector on itself is the square of the length of the vector. That is,

$$\mathbf{A} \cdot \mathbf{A} = |\mathbf{A}|^2$$

In Fig. 9-24 the point M is the foot of the perpendicular to the vector $\mathbf{A}$ drawn from the tip of $\mathbf{B}$. The vector from O to M is called the *vector projection* of $\mathbf{B}$ on $\mathbf{A}$. The vector projection and $\mathbf{A}$ point in the same direction, since θ is an acute angle. If θ exceeds $90°$, then $\mathbf{A}$ and the vector from O to M point oppositely. The *scalar projection* of $\mathbf{B}$ on $\mathbf{A}$ is defined as $|\mathbf{B}| \cos \theta$; the sign of the scalar projection depends on $\cos \theta$. Using the idea of scalar projection of one vector on another, we can interpret the dot product geometrically as

$$\begin{aligned} \mathbf{A} \cdot \mathbf{B} &= |\mathbf{A}| \, |\mathbf{B}| \cos \theta \\ &= \text{(length of } \mathbf{A}) \text{ times (the scalar projection of } \mathbf{B} \text{ on } \mathbf{A}) \end{aligned}$$

We could also say that the dot product of $\mathbf{A}$ and $\mathbf{B}$ is the length of $\mathbf{B}$ times the scalar projection of $\mathbf{A}$ on $\mathbf{B}$.

By definition, $\mathbf{A} \cdot \mathbf{B}$ and $\mathbf{B} \cdot \mathbf{A}$ have exactly the same scalar factors and therefore

$$\mathbf{A} \cdot \mathbf{B} = \mathbf{B} \cdot \mathbf{A} \tag{9}$$

This equation expresses the commutative property for the scalar multiplication of vectors. We next establish the distributive law.

Theorem 9-5   *The scalar product of vectors is distributive. That is*

$$\mathbf{A} \cdot (\mathbf{B} + \mathbf{C}) = \mathbf{A} \cdot \mathbf{B} + \mathbf{A} \cdot \mathbf{C} \tag{10}$$

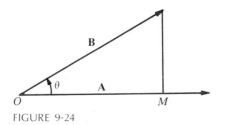

FIGURE 9-24

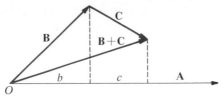

FIGURE 9-25

Proof   If we let b and c stand for the scalar projections of **B** and **C** on **A,** we see (Fig. 9-25) that the sum of the scalar projections of **B** and **C** on **A** is the same as the scalar projection of (**B** + **C**) on **A.** Hence

$$|A|(b + c) = |A|b + |A|c$$

and

$$A \cdot (B + C) = A \cdot B + A \cdot C \tag{10}$$

From equations (9) and (10) we can deduce that the scalar product of sums of vectors may be carried out as in multiplying two algebraic expressions, each of more than one term. Thus, for example,

$$(A + B) \cdot (C + D) = A \cdot (C + D) + B \cdot (C + D)$$
$$= A \cdot C + A \cdot D + B \cdot C + B \cdot D$$

If m and n are scalars, then

$$(mA) \cdot (nB) = mn(A \cdot B) \tag{11}$$

The equation is true if either m or n is equal to zero. But if m and n are both positive or both negative, then mn is positive, and consequently

$$(mA) \cdot (nB) = |mA| \, |nB| \cos \theta$$
$$= mn(A \cdot B)$$

If m and n have opposite signs, the angle between **A** and **B** and the angle between m**A** and n**B** differ by 180°. Then observing Fig. 9-26 and noting that mn is negative, we may write

$$(mA) \cdot (nB) = |mA| \, |nB| \cos (180° - \theta)$$
$$= -|mA| \, |nB| \cos \theta$$
$$= mn(A \cdot B)$$

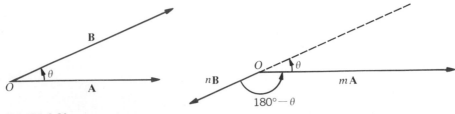

FIGURE 9-26

Theorem 9-6   *If the vectors* **A** *and* **B** *are expressed in terms of the unit vectors* **i** *and* **j** *by*

$$A = a_1i + a_2j$$
$$B = b_1i + b_2j$$

then

$$A \cdot B = a_1b_1 + a_2b_2$$

Proof   We first note that

$$i \cdot i = j \cdot j = 1 \quad \text{and} \quad i \cdot j = j \cdot i = 0$$

Applying these conditions, we have

$$
\begin{aligned}
A \cdot B &= (a_1i + a_2j) \cdot (b_1i + b_2j) \\
&= a_1i \cdot (b_1i + b_2j) + a_2j \cdot (b_1i + b_2j) \\
&= a_1b_1i \cdot i + a_1b_2i \cdot j + a_2b_1j \cdot i + a_2b_2j \cdot j
\end{aligned}
$$

Hence

$$A \cdot B = a_1b_1 + a_2b_2 \tag{12}$$

This equation shows that the dot product is obtained by the simple process of adding the products of the corresponding coefficients of **i** and **j**.

EXAMPLE 1   Determine if the vectors

$$A = 3i + 5j$$
$$B = 5i - 3j$$

are perpendicular.

Solution   The dot product of the vectors, according to formula (12), is

$$A \cdot B = (3)(5) + (5)(-3) = 15 - 15 = 0$$

The dot product is equal to zero, and therefore the vectors are perpendicular.

EXAMPLE 2   The points $P(-2,1)$, $Q(6,-1)$, and $R(2,2)$ are vertices of a triangle. Find the angle θ between the vectors **PQ** and **PR**.

Solution   Let the vector from P to Q be denoted by **A** and the vector from P to R by **B** (Fig. 9-27). Then

$$A = 8i - 2j$$
$$B = 4i + j$$

To find the angle, we substitute in both members of the equation

$$|A| \, |B| \cos \theta = A \cdot B$$

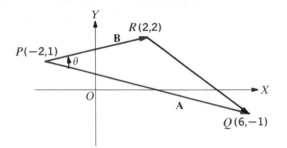

FIGURE 9-27

The lengths of **A** and **B** are $2\sqrt{17}$ and $\sqrt{17}$, and $\mathbf{A} \cdot \mathbf{B} = 30$. Hence

$\cos \theta = \frac{30}{34} = 0.882$

$\theta = \text{Cos}^{-1} 0.882 = 28°$ nearest degree

EXAMPLE 3   Find the scalar projection and the vector projection of **B** on **A** if

$\mathbf{A} = 12\mathbf{i} - 5\mathbf{j}$
$\mathbf{B} = -\mathbf{i} + 4\mathbf{j}$

Solution   The scalar projection of **B** on **A** is $|\mathbf{B}| \cos \theta$. So starting with $|\mathbf{A}| |\mathbf{B}| \cos \theta = \mathbf{A} \cdot \mathbf{B}$, we have

$$|\mathbf{B}| \cos \theta = \frac{\mathbf{A} \cdot \mathbf{B}}{|\mathbf{A}|}$$

$$= \frac{-12 - 20}{\sqrt{144 + 25}}$$

$$= -\frac{32}{13}$$

This negative scalar projection of **B** on **A** means that the given vectors make an angle between 90° and 180°. Consequently the vector projection is directed oppositely to **A**. We obtain the vector projection by multiplying the scalar projection by a unit vector in the direction of **A**. Thus

$$-\frac{32}{13} \frac{12\mathbf{i} - 5\mathbf{j}}{13} = -\frac{32}{169}(12\mathbf{i} - 5\mathbf{j}) = \frac{32}{169}(-12\mathbf{i} + 5\mathbf{j})$$

Exercise 9-4

Find the dot product of the vectors in each problem 1 through 4. Find also the cosine of the angle between the vectors.

1. $A = 3\mathbf{i} + 4\mathbf{j}$ 2. $A = 5\mathbf{i} + 12\mathbf{j}$
   $B = 4\mathbf{i} - 3\mathbf{j}$ $B = 4\mathbf{i} + 3\mathbf{j}$

3. $A = 3i - 5j$
  $B = 6i + 10j$

4. $A = 2i + 8j$
  $B = 4i + j$

5. Find a so that the vectors $i - 3j$ and $2i + aj$ are perpendicular.

6. Find a so that the vectors $-2i + 5j$ and $ai + 3j$ are perpendicular.

The points in problems 7 and 8 are vertices of a triangle. Find the angle θ between the vectors **PQ** and **PR.** Also find the cosine of the remaining angles of the triangle.

7. $P(-2,-2)$, $Q(12,0)$, $R(4,4)$ **8.** $P(4,4)$, $Q(-1,1)$, $R(7,-1)$

In problems 9 through 12 find the scalar projection of **B** on **A.**

9. $A = 4i - 3j$
  $B = 2i + 5j$

10. $A = 5i + 12j$
  $B = 6i + j$

11. $A = 2i + 2j$
  $B = 7i - 6j$

12. $A = -7i + 24j$
  $B = 11i + 4j$

Chapter 9   Review exercise

1. Define: *vector, zero vector, equal vectors.*
2. A body weighing 300 pounds is held in equilibrium (at rest) by two ropes, one making an angle of 32° with the vertical and the other an angle of 42° with the vertical. Find the force exerted on the body by each rope.
3. A force of 28 pounds and a force of 42 pounds act at right angles to each other. Find the magnitude of the resultant and the angle which it makes with each force.
4. Three forces of 36, 40, and 58 pounds acting at a point are in equilibrium. Find to the nearest degree the angles between the directions of the forces.
5. Find the coordinates of the midpoint of the line segment joining the points $P(2,-4)$ and $Q(2,3)$.
6. Find the coordinates of the trisection points of the line segment connecting the points $P(1,4)$ and $Q(5,0)$.

complex numbers

10-1 The imaginary unit

Many equations cannot be solved and many problems cannot be investigated in the real number system. As examples, neither of the equations $x^2 - 4x + 7 = 0$ and $x^2 + 1 = 0$ has a real solution. Clearly the second equation has a solution if and only if $x^2 = -1$. But there is no real number whose square is negative. In view of the inadequacy of real numbers in this and many other situations, we need to enlarge our number system to include a set of new numbers and to define certain operations on them. The use of the new numbers has contributed to the development of a large body of mathematics, much of which has vital application in the engineering and physical sciences.

As a first step in extending the real number system, we introduce the number $\sqrt{-1}$, called the *imaginary unit,* whose square is defined to be -1. The imaginary unit is customarily denoted by i; that is, $i = \sqrt{-1}$. Also we agree to express the product of i and a real number b by bi and the sum of a real number a and bi by $a + bi$.

Now that we have introduced a square root of -1 into our system of numbers, let us consider the square roots of any negative number $-p$. Since $i^2 = -1$, we may write

$$(i\sqrt{p})^2 = i^2 p = -p \qquad \text{and} \qquad (-i\sqrt{p})^2 = i^2 p = -p$$

Hence the square roots of $-p$ are $i\sqrt{p}$ and $-i\sqrt{p}$. To distinguish between these roots, we let

$$\sqrt{-p} = i\sqrt{p} \qquad \text{and} \qquad -\sqrt{-p} = -i\sqrt{p}$$

As examples, we have

$$\sqrt{-5} = i\sqrt{5}$$
$$\sqrt{-9} = i\sqrt{9} = 3i$$
$$-\sqrt{-4} = -i\sqrt{4} = -2i$$

By employing $i^2 = -1$, we see that

$$i^3 = i^2 \cdot i = -i$$
$$i^4 = i^2 \cdot i^2 = (-1)(-1) = 1$$

Since $i^4 = 1$, it follows that i with an exponent which is an integral multiple of 4 is also equal to 1. Consequently, integral powers of i yield only the numbers i, -1, $-i$, and 1. Thus,

$$i^{13} = i^{12} \cdot i = i$$
$$i^{-6} = i^{-6} \cdot i^8 = i^2 = -1$$

10-2   Complex numbers

Having enlarged the number system by introducing the imaginary unit i and numbers of the form bi, b a real number, we now find it useful to make a further enlargement.

Definition 10-1   *A number of the form $a + bi$, with a and b real constants, is called a* complex number. *The number a is called the* real component *and b is called the* imaginary component.

Depending on the values of a and b, we have the following special cases of a complex number.

Definition 10-2   *$a + bi$ is called an* imaginary number *if $b \neq 0$.*
$a + bi$ is called a pure imaginary number *if $a = 0$ and $b \neq 0$.*
$a + bi$ is called a real number *if $b = 0$.*

EXAMPLES $2i$, $3 - 5i$, and $-4 + i$ are imaginary numbers.
$2i$, $-6i$, and $i\sqrt{5}$ are pure imaginary numbers.
7, -2.5, and 0 are real numbers.

The set of numbers determined by $x + yi$, where x and y are real numbers, comprises the *complex number system.* Any real number x and the complex number $x + 0i$ are considered to be the same, and if y is a real number, the complex number $0 + yi$ and the pure imaginary number yi are also considered to be the same. Hence, as examples, we write

$$2 + 0i = 2 \quad \text{and} \quad 0 - 5i = -5i$$

We conclude, then, that the set of real numbers and the set of pure imaginary numbers are special parts (called subsets) of the system of complex numbers.

Definition 10-3   *Two complex numbers are equal if and only if the real compo-nents are equal and the imaginary components are equal. That is,*

$$a + bi = c + di \qquad \textit{if and only if } a = c \textit{ and } b = d$$

EXAMPLE   If $2x - 5 + (4x - 1)i = y + 3 + (3x - y)i$, find the values of x and y.

Solution   The real components of the members of the equation must be equal and the imaginary components must be equal. Hence we have the system of equations

$$2x - 5 = y + 3$$
$$4x - 1 = 3x - y$$

The solution of this system is $x = 3, y = -2$.

10-3   Fundamental operations on complex numbers

In this section we consider the operations of addition, multiplication, sub-traction, and division of complex numbers.

Definition 10-4   *The sum, product, difference, and quotient of two complex numbers are given, respectively, by the following equations:*

sum: $(a + bi) + (c + di) = (a + c) + (b + d)i$

product: $(a + bi)(c + di) = (ac - bd) + (bc + ad)i$

difference: $(a + bi) - (c + di) = (a - c) + (b - d)i$

quotient: $\dfrac{a + bi}{c + di} = \dfrac{(ac + bd) + (bc - ad)i}{c^2 + d^2}$

We note that the sum and product are obtained by carrying out the operations as though i were a real number and, for the product, by replacing i^2 with -1. We obtain the difference by adding the negative of the sub-trahend. Thus

$$(a + bi) - (c + di) = (a + bi) + (-c - di)$$
$$= (a - c) + (b - d)i$$

Finally we multiply the dividend and divisor by $c - di$ to get the quotient.

$$\frac{a + bi}{c + di} = \frac{(a + bi)(c - di)}{(c + di)(c - di)} = \frac{(ac + bd) + (bc - ad)i}{c^2 + d^2}$$

The numbers $c + di$ and $c - di$ are said to be *conjugate complex numbers*. Each is the conjugate of the other. Thus $4 + 3i$, $4 - 3i$, and $5i$, $-5i$ are pairs of conjugate complex numbers.

If a, b, and c are any real numbers, the following equations are assumed to be true.

commutative law: $a + b = b + a$ $ab = ba$

associative law: $(a + b) + c = a + (b + c)$ $(ab)c = a(bc)$

distributive law: $a(b + c) = ab + bc$

By using Definition 10-4 and the preceding laws for real numbers, we are able to establish the corresponding laws for complex numbers. We shall, however, omit the proofs of these properties except for the commutative and associative laws for addition.

Theorem 10-1   *The commutative and associative laws are valid for the addition of complex numbers.*

Proof

$$(a + bi) + (c + di) = (a + c) + (b + d)i \qquad \text{by definition}$$
$$= (c + a) + (d + b)i \qquad \text{by the commutative law}$$
$$= (c + di) + (a + bi) \qquad \text{by definition}$$

Next, using the definition of addition of complex numbers and the associative law for the addition of real numbers, we have

$$[(a + bi) + (c + di)] + (e + fi) = [(a + c) + (b + d)i] + (e + fi)$$
$$= [(a + c) + e] + [(b + d) + f]i$$
$$= [a + (c + e)] + [b + (d + f)]i$$
$$= (a + bi) + [(c + di) + (e + fi)]$$

The commutative and associative properties for the multiplication of complex numbers, as well as the distributive property, can be similarly established. The associative property, however, can be more conveniently proved after we have considered the multiplication of numbers in polar form (Sec. 10-6).

EXAMPLE 1   Find the product of $\sqrt{-5}$ and $\sqrt{-7}$.

Solution   We first express each number as a real number times i and then multiply. Thus

$$\sqrt{-5} \cdot \sqrt{-7} = i\sqrt{5} \cdot i\sqrt{7} = i^2\sqrt{35} = -\sqrt{35}$$

This example illustrates the fact that the law for multiplying two radicals of like order is not applicable for imaginary numbers. Although $\sqrt{5} \cdot \sqrt{7} = \sqrt{35}$, the product $\sqrt{-5} \cdot \sqrt{-7} \neq \sqrt{35}$. An error is less likely to be made in a product of this kind if each factor is expressed as a real number times i before multiplying.

EXAMPLE 2   Find the quotient of $2 - \sqrt{-9}$ divided by $4 + \sqrt{-25}$.

Solution   We express these numbers as $2 - 3i$ and $4 + 5i$. Then, remembering that $i^2 = -1$, we get

$$\frac{2 - 3i}{4 + 5i} = \frac{(2 - 3i)(4 - 5i)}{(4 + 5i)(4 - 5i)}$$

$$= \frac{8 + 15i^2 - 12i - 10i}{16 - 25i^2}$$

$$= \frac{-7 - 22i}{41}$$

In practice, the method used in Example 2 for finding the quotient of two complex numbers is more convenient than using the quotient formula.

Exercise 10-1

Simplify the following expressions.

1. $3i^7 - 2i^4 + 4i^6$ **2.** $5i^{11} + 3i^3 + 6i^5$
3. $2i^2 + 5i^{13} - 6i^{23}$ **4.** $9i^{10} + 8i^{11} - 7i^{12}$
5. $4i^6 + 9i^{13} - 5i^{14}$ **6.** $3i^4 - 2i^{15} + 6i^{19}$

Solve each of the following equations for x and y.

7. $3 - ix = y + 4i$ **8.** $ix - 2y = 3 + 4i$
9. $3y - 2 + (4x + 7)i = 0$ **10.** $2x - (y + x)i = 5$
11. $x - y + i = 3 + (y - 2x)i$ **12.** $6x + (x - 2y)i = 3y - i$

Perform the indicated operations, leaving each result in the form $a + bi$.

13. $(3 + 4i) + (7 + 5i)$ **14.** $(3 - i) + (-4 + 5i)$
15. $(2 - 3i) - (4 + 6i)$ **16.** $(3 + 9i) - (4 - 7i)$
17. $(4 - 4i) - (-3 - 4i)$ **18.** $(7 - 2i) - (8 - 3i)$
19. $(i - 3) + (-1 + i) + 4i$ **20.** $(4 - 6i) + (3 - 9i) + 4$
21. $4\sqrt{-9} + 2\sqrt{-4} - 6\sqrt{-25} + 3\sqrt{-36}$
22. $(4 + \sqrt{-81}) + (5 - 2\sqrt{-49}) - (1 + \sqrt{-1})$
23. $6\sqrt{-8} + 5\sqrt{-3} - 3\sqrt{-18} - \sqrt{-32}$

24. $(2 + \sqrt{-12} - (1 + \sqrt{-27}) + (3 + \sqrt{-48})$
25. $(3 - i)(5 + 4i)$ 26. $(4 + 3i)(5 + 2i)$
27. $(6 - 4i)(6 + 4i)$ 28. $(5 + 4i)(5 - 4i)$
29. $(8 - 2i)(1 - 3i)$ 30. $(2 + 5i)(5 + 2i)(2 - i)$
31. $(2 + i)(1 - i)$ 32. $(2 - 4i)(4 + 2i)(1 + i)$

33. $\dfrac{4 + i}{i}$ 34. $\dfrac{12 - 6i}{3i}$ 35. $\dfrac{3}{3 - 4i}$

36. $\dfrac{8i}{2 + i}$ 37. $\dfrac{2 + i}{6 - i}$ 38. $\dfrac{3 - 2i}{3 + i}$

39. $\dfrac{2 - 3i}{17 - 6i}$ 40. $\dfrac{2 - 3i}{2 + 3i}$ 41. $\dfrac{9 + 7i}{9 - 7i}$

42. Prove that the sum and product of two conjugate complex numbers are both real.
43. Prove that $x + yi = 0$ if and only if $x = 0$ and $y = 0$.
44. Prove that multiplication of two complex numbers is commutative.
45. If the product of two complex numbers is zero, prove that at least one of the factors is zero.
46. Prove that the conjugate of the sum of two complex numbers is equal to the sum of their conjugates.

10-4   Graphical representation of complex numbers

Complex numbers are usually represented by points in the plane of a rectangular coordinate system. The number $a + bi$ is represented by the point whose coordinates are (a,b). Hence any complex number determines a point in the plane. Conversely, any point of the plane determines a complex number. The points on the axes correspond to special complex numbers. Points on the x axis represent real numbers because the ordinates of the points are zero. Similarly, the points on the y axis have abscissas equal to zero and therefore correspond to pure imaginary numbers. Thus the point $(a,0)$ corresponds to the real number $a = a + 0i$, and the point $(0,b)$ corresponds to the pure imaginary number $bi = 0 + bi$.

In Fig. 10-1, the correspondence between some complex numbers and points is indicated.

When a rectangular coordinate plane is used to represent complex numbers, the x axis is called the *real axis*, the y axis the *pure imaginary axis,* and the plane the *complex plane.*

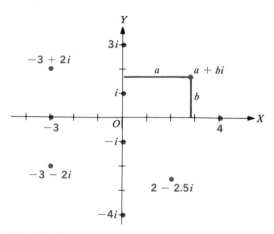

FIGURE 10-1

The sum of two complex numbers may be found graphically by constructing a parallelogram. The two points representing the given numbers are opposite vertices, and the origin is a third vertex. The fourth vertex of the parallelogram thus determined represents the sum. This method of adding may be verified by observing that the opposite sides of the quadrilateral (Fig. 10-2) are parallel. Further, since the coordinates of the vertices are $(0,0)$, (a,b), (c,d), and $(a + c, b + d)$, the distance formula (Sec. 1-5) reveals that $\sqrt{a^2 + b^2}$ is the length of each of one pair of opposite sides and $\sqrt{c^2 + d^2}$ is the length of each of the other sides.

The parallelogram method of adding two complex numbers is essentially the method of adding two vectors (Sec. 9-2). This suggests that a complex number represented by a point may also be represented by a vector drawn from the origin to the point. Often complex numbers are so represented. Figure 10-3 shows the vectors for the numbers $8 + 2i$ and $-4 + 3i$ and the vector for their sum.

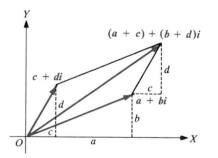

FIGURE 10-2

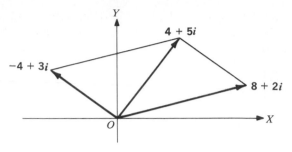

Y

4 + 5i

−4 + 3i

8 + 2i

O

X

FIGURE 10-3

10-5 Trigonometric form of a complex number

A complex number expressed as $a + bi$ is said to be in *algebraic,* or *rectangular, form.* We next show how to express a complex number in another useful form.

The point P (Fig. 10-4) corresponding to the number $a + bi$ has the coordinates (a,b). The distance from the origin to P is denoted by r, and the angle from the positive real axis to the distance segment is denoted by θ. The distance r (chosen positive when P is any point other than the origin) is called the *absolute value,* or *modulus.* The angle θ is called the *amplitude,* or *argument.*

Referring to the diagram, we have the relations

$$r = \sqrt{a^2 + b^2} \qquad \tan \theta = \frac{b}{a}$$

and

$$a = r \cos \theta \qquad b = r \sin \theta$$

Hence it follows that

$$a + bi = r(\cos \theta + i \sin \theta)$$

The expression $r(\cos \theta + i \sin \theta)$ is called the *trigonometric,* or *polar, form* of a complex number. The number r is never negative since it is the modulus. The factor $(\cos \theta + i \sin \theta)$ reveals the direction of the point $P(a + bi)$ from

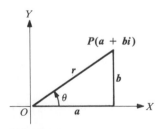

Y

P(a + bi)

r

b

θ

O

a

X

FIGURE 10-4

the origin. This is true because θ must satisfy the equations $r\cos\theta = a$ and $r\sin\theta = b$, with r positive. The preceding formulas, then, are sufficient for changing a number from algebraic form to trigonometric form and vice versa.

EXAMPLE 1   Plot the number $2\sqrt{3} - 2i$ and change to the trigonometric form.

Solution   The point $(2\sqrt{3}, -2)$ represents the number (Fig. 10-5). Letting $a = 2\sqrt{3}$ and $b = -2$ in the preceding formulas, we obtain

$$r = \sqrt{12 + 4} = 4 \qquad \cos\theta = \frac{\sqrt{3}}{2} \qquad \sin\theta = -\tfrac{1}{2}$$

We choose the smallest positive value of θ which satisfies the last two equations and have

$$2\sqrt{3} - 2i = 4(\cos 330° + i\sin 330°)$$

EXAMPLE 2   Plot the number $5(\cos 120° + i\sin 120°)$ and change to algebraic form.

Solution   The point corresponding to this number is located by drawing the angle 120° in standard position and measuring 5 units off along the terminal side. From Fig. 10-6, we have

$$a = 5\cos 120° = 5(-\tfrac{1}{2}) = -\tfrac{5}{2}$$

$$b = 5\sin 120° = 5\left(\frac{\sqrt{3}}{2}\right) = \frac{5\sqrt{3}}{2}$$

and therefore

$$5(\cos 120° + i\sin 120°) = -\frac{5}{2} + \frac{5\sqrt{3}}{2}i$$

EXAMPLE 3   Express the number $4 + 3i$ in trigonometric form.

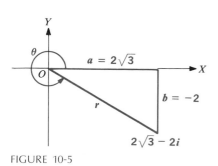

FIGURE 10-5

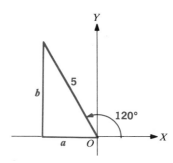

FIGURE 10-6

Solution   Since $a = 4$ and $b = 3$, we have

$$r = \sqrt{16 + 9} = 5 \qquad \cos \theta = \tfrac{4}{5} \qquad \sin \theta = \tfrac{3}{5}$$

Referring to a table, we find $\theta = 36° 52'$, and hence

$$4 + 3i = 5(\cos 36° 52' + i \sin 36° 52')$$

Exercise 10-2

Write the conjugate of each complex number. Plot the number and its conjugate.

1. $3 - 2i$	**2.** $6 + 5i$	**3.** $0 + 3i$
4. $-5 + 0i$	**5.** $-4 + 2i$	**6.** $4 - 7i$
7. $-3 - 4i$	**8.** $4 - 3i$	**9.** $7 + 4i$

Plot each complex number and write the corresponding trigonometric form.

10. $1 + i$	**11.** $-1 - i$	**12.** $3 + 0i$
13. $\sqrt{3} - i$	**14.** $2\sqrt{3} + 2i$	**15.** $1 - i\sqrt{3}$
16. $0 - 2i$	**17.** $-3 + 0i$	**18.** $0 + 5i$
19. $\sqrt{2} + i\sqrt{2}$	**20.** $5 - 5i\sqrt{3}$	**21.** $-\sqrt{15} - i\sqrt{5}$
22. $\sqrt{7} - i\sqrt{21}$	**23.** $4 + 3i$	**24.** $2 - i$

Change each number to algebraic form.

25. $2(\cos 60° + i \sin 60°)$	**26.** $4(\cos 30° + i \sin 30°)$
27. $2(\cos 45° + i \sin 45°)$	**28.** $8(\cos 135° + i \sin 135°)$
29. $4(\cos 150° + i \sin 150°)$	**30.** $\cos 90° + i \sin 90°$
31. $3(\cos 180° + i \sin 180°)$	**32.** $6(\cos 270° + i \sin 270°)$
33. $\cos 77° + i \sin 77°$	**34.** $8(\cos 228° + i \sin 228°)$

Plot each pair of numbers and their sum. Draw the parallelogram determined by the two points and the origin.

35. $2 + 3i, 3 + i$	**36.** $4 - 2i, 2 - 5i$	**37.** $3 + 2i, 2 + 3i$
38. $1 + 4i, 1 - 2i$	**39.** $5 + 3i, 5 - 3i$	**40.** $6 + 2i, 2 - 6i$

10-6   Multiplication and division of numbers in polar form

The product and quotient of two complex numbers may be written immediately if the numbers are in polar form.

Theorem 10-2   *The absolute value of the product of two complex numbers is equal to the product of their absolute values. The amplitude of the product of two complex numbers is equal to the sum of their amplitudes.*

Proof   Let the two complex numbers be denoted by

$$z_1 = r_1(\cos \theta_1 + i \sin \theta_1)$$

and

$$z_2 = r_2(\cos \theta_2 + i \sin \theta_2)$$

Then, multiplying in the usual way, we obtain

$$z_1 z_2 = r_1(\cos \theta_1 + i \sin \theta_1) \cdot r_2(\cos \theta_2 + i \sin \theta_2)$$
$$= r_1 r_2[(\cos \theta_1 \cos \theta_2 + i^2 \sin \theta_1 \sin \theta_2)$$
$$+ i(\sin \theta_1 \cos \theta_2 + \sin \theta_2 \cos \theta_1)]$$

Replacing i^2 by -1 and applying identities (17) and (23) of Sec. 4-10 for the cosine and the sine of the sum of two angles, we get

$$r_1(\cos \theta_1 + i \sin \theta_1) \cdot r_2(\cos \theta_2 + i \sin \theta_2)$$
$$= r_1 r_2[\cos (\theta_1 + \theta_2) + i \sin (\theta_1 + \theta_2)]$$

This law of multiplication may be applied repeatedly to give the product of three or more complex numbers. For the three numbers

$$z_1 = r_1(\cos \theta_1 + i \sin \theta_1)$$
$$z_2 = r_2(\cos \theta_2 + i \sin \theta_2)$$
$$z_3 = r_3(\cos \theta_3 + i \sin \theta_3)$$

we have

$$z_1 z_2 = r_1 r_2[\cos (\theta_1 + \theta_2) + i \sin (\theta_1 + \theta_2)]$$

and then

$$z_1 z_2 z_3 = r_1 r_2 r_3[\cos (\theta_1 + \theta_2 + \theta_3) + i \sin (\theta_1 + \theta_2 + \theta_3)]$$

The product of n complex numbers is given by

$$z_1 z_2 \cdots z_n = r_1 r_2 \cdots r_n[\cos (\theta_1 + \theta_2 + \cdots + \theta_n)$$
$$+ i \sin (\theta_1 + \theta_2 + \cdots + \theta_n)]$$

We suggest that the student prove the associative property of complex numbers for multiplication; that is, prove $(z_1 z_2)z_3 = z_1(z_2 z_3)$.

Theorem 10-3   *The absolute value of the quotient of two complex numbers is the quotient of their absolute values. The amplitude is the amplitude of the dividend minus the amplitude of the divisor.*

Proof   To find the quotient of z_1 divided by z_2, we multiply the dividend and divisor by $\cos\theta_2 - i\sin\theta_2$. Thus we have

$$\frac{z_1}{z_2} = \frac{r_1(\cos\theta_1 + i\sin\theta_1)(\cos\theta_2 - i\sin\theta_2)}{r_2(\cos\theta_2 + i\sin\theta_2)(\cos\theta_2 - i\sin\theta_2)}$$

The product of the factors in the numerator is

$$r_1[(\cos\theta_1\cos\theta_2 + \sin\theta_1\sin\theta_2) + i(\sin\theta_1\cos\theta_2 - \cos\theta_1\sin\theta_2)]$$

and the product of the factors of the denominator is

$$r_2(\cos^2\theta_2 - i^2\sin^2\theta_2) = r_2$$

We now apply identities (16) and (24) of Sec. 4-10 for the cosine and sine of the difference of two angles and express the numerator in the form $r_1[\cos(\theta_1 - \theta_2) + i\sin(\theta_1 - \theta_2)]$. Then we obtain the formula

$$\frac{r_1(\cos\theta_1 + i\sin\theta_1)}{r_2(\cos\theta_2 + i\sin\theta_2)} = \frac{r_1}{r_2}[\cos(\theta_1 - \theta_2) + i\sin(\theta_1 - \theta_2)]$$

The two preceding theorems may be applied graphically to give the product and quotient of two complex numbers. To find the product of $r_1(\cos\theta_1 + i\sin\theta_1)$ and $r_2(\cos\theta_2 + i\sin\theta_2)$, draw the angle $(\theta_1 + \theta_2)$ in standard position and measure off a distance on the terminal side equal to $r_1 r_2$. To find the quotient of $r_1(\cos\theta_1 + i\sin\theta_1)$ divided by $r_2(\cos\theta_2 + i\sin\theta_2)$, draw the angle $(\theta_1 - \theta_2)$ in standard position and measure off a distance along the terminal side equal to r_1/r_2.

EXAMPLE 1   Express the numbers $1 + i\sqrt{3}$ and $3 - 3i$ in polar forms and find their product.

Solution   We have for the polar forms

$$1 + i\sqrt{3} = 2(\cos 60° + i\sin 60°)$$

and

$$3 - 3i = 3\sqrt{2}(\cos 315° + i\sin 315°)$$

Hence,

$$(1 + i\sqrt{3})(3 - 3i) = 6\sqrt{2}(\cos 375° + i\sin 375°)$$
$$= 6\sqrt{2}(\cos 15° + i\sin 15°)$$

EXAMPLE 2   Find the quotient of $12(\cos 30° + i\sin 30°)$ divided by $3(\cos 70° + i\sin 70°)$.

Solution   Applying the law for division, we obtain

$$\frac{12(\cos 30° + i\sin 30°)}{3(\cos 70° + i\sin 70°)} = 4[\cos(-40°) + i\sin(-40°)]$$
$$= 4(\cos 320° + i\sin 320°)$$

Exercise 10-3

Perform the indicated operations, leaving each result in the form $a + bi$.

1. $2(\cos 40° + i \sin 40°) \cdot 3(\cos 80° + i \sin 80°)$
2. $4(\cos 100° + i \sin 100°) \cdot 5(\cos 80° + i \sin 80°)$
3. $6(\cos 45° + i \sin 45°) \cdot 3(\cos 75° + i \sin 75°)$
4. $2(\cos 125° + i \sin 125°) \cdot 7(\cos 100° + i \sin 100°)$
5. $3(\cos 170° + i \sin 170°) \cdot 9(\cos 280° + i \sin 280°)$

6. $\dfrac{\cos 120° + i \sin 120°}{\cos 30° + i \sin 30°}$ 7. $\dfrac{8(\cos 264° + i \sin 264°)}{4(\cos 84° + i \sin 84°)}$

8. $\dfrac{9(\cos 310° + i \sin 310°)}{3(\cos 265° + i \sin 265°)}$ 9. $\dfrac{14(\cos 31° + i \sin 31°)}{2(\cos 91° + i \sin 91°)}$

Express each number in trigonometric form and then perform the indicated operations. Leave the result in trigonometric form.

10. $(1 + i)(1 - i\sqrt{3})$ 11. $(\sqrt{3} - i)(1 - i)$
12. $(2 - 2i)(3 - 3i)$ 13. $5i(4 + 4i)$
14. $2i(1 - i\sqrt{3})(\sqrt{3} - i)$ 15. $(3 + 3i)(-\sqrt{3} + i)(1 + i\sqrt{3})$

16. $\dfrac{2}{3 - 3i}$ 17. $\dfrac{2\sqrt{3} + 2i}{-1 + i\sqrt{3}}$ 18. $\dfrac{4}{3\sqrt{3} + 3i}$ 19. $\dfrac{1 + i}{1 - i\sqrt{3}}$

Perform graphically the indicated operations.

20. $3(\cos 45° + i \sin 45°) \cdot 2(\cos 45° + i \sin 45°)$
21. $3(\cos 30° + i \sin 30°) \cdot (\cos 90° + i \sin 90°)$
22. $2(\cos 60° + i \sin 60°) \cdot 4(\cos 90° + i \sin 90°)$

23. $\dfrac{6(\cos 225° + i \sin 225°)}{3(\cos 90° + i \sin 90°)}$ 24. $\dfrac{10(\cos 270° + i \sin 270°)}{5(\cos 135° + i \sin 135°)}$

10-7   Powers and roots of numbers

In this section we shall be interested in finding positive integral powers of complex numbers and roots of complex numbers. The following theorem opens the way for the performance of these operations.

Theorem 10-4   *If n is any real number, the nth power of the complex number $r(\cos \theta + i \sin \theta)$ is given by the formula*

$$[r(\cos \theta + i \sin \theta)]^n = r^n(\cos n\theta + i \sin n\theta) \tag{1}$$

We shall not prove this theorem. We point out, however, that the theorem may be established at once for the case in which n is a positive integer. This is accomplished simply by letting each of the factors in the formula for the product of n complex numbers be equal to $r(\cos \theta + i \sin \theta)$. When n is a positive integer the preceding theorem is called *De Moivre's theorem* in honor of the French mathematician Abraham De Moivre.

Formula (1) yields readily a positive integral power of a complex number. To find roots of numbers, however, we convert the formula to a more useful form.

The number $s(\cos \alpha + i \sin \alpha)$ is, by definition, an nth root of $r(\cos \theta + i \sin \theta)$ if

$$[s(\cos \alpha + i \sin \alpha)]^n = r(\cos \theta + i \sin \theta)$$

We apply De Moivre's formula (1) to the left member of this equation and, in the right member, replace θ by $\theta + k \cdot 360°$, where k is any integer. This replacement is justified by the fact that the values of the trigonometric functions of θ and $\theta + k \cdot 360°$ are the same (Sec. 2-4). Thus we have

$$s^n(\cos n\alpha + i \sin n\alpha) = r[\cos (\theta + k \cdot 360°) + i \sin (\theta + k \cdot 360°)]$$

We seek values of s and α so that this equation will be satisfied. Consequently, we equate the moduli and amplitudes of the members of the equation. This gives

$$s^n = r \qquad \text{and} \qquad n\alpha = \theta + k \cdot 360°$$

from which, solving for s and α, we have

$$s = \sqrt[n]{r} \qquad \text{and} \qquad \alpha = \frac{\theta + k \cdot 360°}{n}$$

Hence

$$s(\cos \alpha + i \sin \alpha) = \sqrt[n]{r} \left(\cos \frac{\theta + k \cdot 360°}{n} + i \sin \frac{\theta + k \cdot 360°}{n} \right)$$

The right member of this equation may be made to take n distinct values by giving k the values $0, 1, 2, \ldots, n - 1$. Any other integral value of k yields a repetition and not a new root. Thus there are n different nth roots of any nonzero number. And it may be observed that the amplitudes are such that graphically the roots are equally spaced about the circumference of a circle of radius $\sqrt[n]{r}$. Hence we have the following theorem.

Theorem 10-5   *A nonzero number $r(\cos \theta + i \sin \theta)$ has n nth roots which are given by the formula*

$$\sqrt[n]{r} \left(\cos \frac{\theta + k \cdot 360°}{n} + i \sin \frac{\theta + k \cdot 360°}{n} \right) \qquad (2)$$

where $k = 0, 1, 2 \ldots, n - 1$.

EXAMPLE 1   Find the fourth power of $(1 - i\sqrt{3})$.

Solution   First changing to polar form, we obtain

$$1 - i\sqrt{3} = 2(\cos 300° + i \sin 300°)$$

Applying formula (1) with $n = 4$, we get

$$
\begin{aligned}
(1 - i\sqrt{3})^4 &= [2(\cos 300° + i \sin 300°)]^4 \\
&= 2^4(\cos 1200° + i \sin 1200°) \\
&= 16(\cos 120° + i \sin 120°) \\
&= -8 + 8i\sqrt{3}
\end{aligned}
$$

EXAMPLE 2   Find the cube roots of -8.

Solution   We first express -8 in polar form. Thus

$$-8 = 8[\cos(180° + k \cdot 360°) + i \sin(180° + k \cdot 360°)]$$

We now apply formula (2) for roots of a number and have

$$
\begin{aligned}
\sqrt[3]{8}\left(\cos \frac{180° + k \cdot 360°}{3} + i \sin \frac{180° + k \cdot 360°}{3}\right) \\
= 2[\cos(60° + k \cdot 120°) + i \sin(60° + k \cdot 120°)]
\end{aligned}
$$

Assigning k the values 0, 1, and 2, in succession, we find the three cube roots of -8 to be

$$
\begin{aligned}
2(\cos 60° + i \sin 60°) &= 1 + i\sqrt{3} \\
2(\cos 180° + i \sin 180°) &= -2 \\
2(\cos 300° + i \sin 300°) &= 1 - i\sqrt{3}
\end{aligned}
$$

The numbers $1 + i\sqrt{3}$, -2, and $1 - i\sqrt{3}$ are the rectangular forms of the cube roots of -8. These are all the cube roots of -8; any other integral value of k gives a repetition of one of these roots. In particular, if the values 3, 4, and 5 are assigned to k, the roots are repeated in the order just mentioned.

EXAMPLE 3   Find the five fifth roots of $-16 + 16i\sqrt{3}$.

Solution   The polar form of $-16 + 16i\sqrt{3}$ is

$$32[\cos(120° + k \cdot 360°) + i \sin(120° + k \cdot 360°)]$$

From formula (2), we write

$$
\begin{aligned}
\sqrt[5]{32}\left(\cos \frac{120° + k \cdot 360°}{5} + i \sin \frac{120° + k \cdot 360°}{5}\right) \\
= 2[\cos(24° + k \cdot 72°) + i \sin(24° + k \cdot 72°)]
\end{aligned}
$$

The desired roots are obtainable by assigning k the values 0, 1, 2, 3, 4,

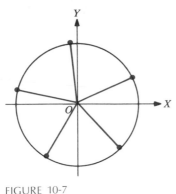

FIGURE 10-7

in succession, in the right member of this equation. Thus the five fifth roots of $-16 + 16i\sqrt{3}$ are

$2(\cos 24° + i \sin 24°)$
$2(\cos 96° + i \sin 96°)$
$2(\cos 168° + i \sin 168°)$
$2(\cos 240° + i \sin 240°)$
$2(\cos 312° + i \sin 312°)$

The points representing these roots are equally spaced about the circle of radius 2 in Fig. 10-7.

Exercise 10-4

Find the indicated powers in the following problems. Express each result in trigonometric form.

1. $[3(\cos 34° + i \sin 34°)]^2$ **2.** $[2(\cos 25° + i \sin 25°)]^3$
3. $[5(\cos 130° + i \sin 130°)]^3$ **4.** $[4(\cos 202° + i \sin 202°)]^4$
5. $(\cos 70° + i \sin 70°)^8$ **6.** $(\cos 100° + i \sin 100°)^9$
7. $(1 - i)^4$ **8.** $(1 + i)^6$ **9.** $(1 - i\sqrt{3})^5$
10. $(\sqrt{3} + i)^6$ **11.** $(2\sqrt{3} - 2i)^5$ **12.** $(2 - 2i)^7$

Find the indicated roots of the given expressions. Leave each result in trigonometric form.

13. The square roots of $9(\cos 40° + i \sin 40°)$
14. The square roots of $25(\cos 132° + i \sin 132°)$
15. The cube roots of $27(\cos 132° + i \sin 132°)$
16. The fourth roots of $16(\cos 148° + i \sin 148°)$
17. The fourth roots of $81(\cos 88° + i \sin 88°)$
18. The fifth roots of $\cos 175° + i \sin 175°$

19. The cube roots of $1 - i$

20. The cube roots of i

21. The fourth roots of $8i$

22. The fourth roots of 32

23. The fifth roots of i

24. The fifth roots of $1 - i$

Find the indicated powers in the following problems. Give each result in algebraic form.

25. $[2(\cos 35° + i \sin 35°)]^6$

26. $[2(\cos 36° + i \sin 36°)]^5$

27. $(\cos 140° + i \sin 140°)^5$

28. $(\cos 130° + i \sin 130°)^6$

29. $(\cos 15° + i \sin 15°)^8$

30. $(\cos 24° + i \sin 24°)^{10}$

Solve the following equations for x. Leave results in algebraic form.

31. $x^3 + i = 0$ **32.** $x^3 - 8i = 0$ **33.** $x^4 + 16 = 0$

34. $x^4 + 32 = 0$ **35.** $x^5 - i = 0$ **36.** $x^6 + 1 = 0$

10-8 Polar coordinates

In the rectangular coordinate system we locate a point by its directed distances from two perpendicular lines. We propose now to introduce a coordinate system in which the coordinates of a point in a plane are its distance from a fixed point and a direction from a fixed line. In Sec. 10-5, we really determined points in this manner. The angle θ (Fig. 10-4) gives the direction from a horizontal line, and r gives the distance from a point of the line. There, however, we associated a point with a complex number. But now we shall drop the idea of complex numbers and simply establish a correspondence between a pair of quantities (coordinates) and a point determined by the coordinates.

In Fig. 10-8, we let OA be a half line. The point O is the origin, or pole, and OA is called the *polar axis*. A point P in a plane containing the polar axis is definitely determined by angle AOP and the distance OP. The angle AOP, denoted by θ, is called the *vectorial angle*, and the segment OP, denoted

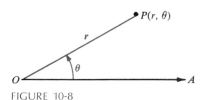

FIGURE 10-8

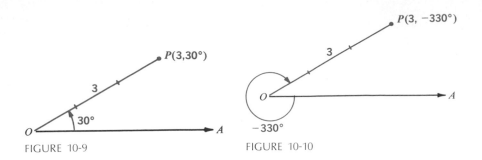

P(3,30°)

3

30°

O   A

FIGURE 10-9

P(3, −330°)

3

O   A

−330°

FIGURE 10-10

by r, is called the *radius vector*. We let the polar axis be the initial side of θ and the half line from O through P the terminal side. The polar coordinates of the point P are denoted by (r,θ).

Polar coordinates are customarily regarded as signed quantities. The vectorial angle is defined as positive if it is measured counterclockwise from the polar axis and negative if measured clockwise from the polar axis. The r coordinate is defined as positive if it is measured from the origin along the terminal side of θ and negative if it is measured from the origin along the terminal side extended through the origin.

A given pair of polar coordinates definitely determines the position of a point. For example, the coordinates $(3,30°)$ determine one particular point. We locate the point by first drawing the terminal side of an angle of $30°$ measured counterclockwise from OA and then measuring a distance of three units along the terminal side (Fig. 10-9). Although this pair of coordinates fixes the position of a point, there are other coordinates which determine the same point. This is evidently true because the vectorial angle may be changed by any integral multiple of $360°$ without changing the point represented. Other coordinates of a point may be found by allowing the r coordinate to be negative. Thus, restricting the vectorial angle to measures from $−330°$ to $210°$, we see (Figs. 10-9 to 10-12) that the following pairs of coordinates determine the same point:

$$(3,30°) \quad (3,−330°) \quad (−3,210°) \quad (−3,−150°)$$

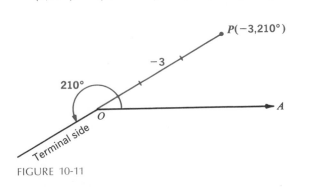

P(−3,210°)

−3

210°

O   A

Terminal side

FIGURE 10-11

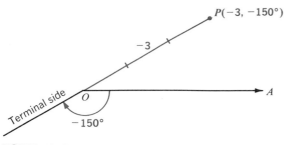

FIGURE 10-12

Although a point of given coordinates can be plotted by estimating the vectorial angle and the distance by sight, greatly improved accuracy may be obtained by the use of polar coordinate paper. This paper has equally spaced circles with their centers at the pole and equally spaced radial lines through the pole (Fig. 10-13). The absolute value of the r coordinate is the same at all points of each circle. In the figure the terminal sides of angles are drawn in steps of $30°$ from $0°$ to $360°$. The r coordinates are positive at all points on the terminal side of an angle and negative on the terminal side extended through the origin, as the plotted points illustrate. Since the r coordinate is equal to zero only at the pole, we may represent this point by $(0,\theta)$, where θ is any angle.

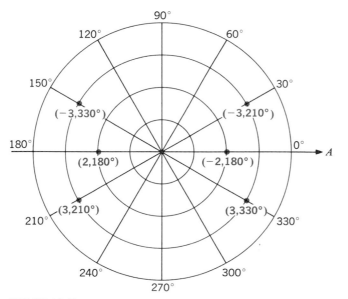

FIGURE 10-13

Exercise 10-5

Plot the point corresponding to each pair of coordinates. Write three other pairs of polar coordinates of the same point, restricting θ so that $-360° < \theta \leq 360°$.

1. $(3,60°)$	**2.** $(4,120°)$	**3.** $(5,0°)$
4. $(2,240°)$	**5.** $(2,90°)$	**6.** $(-1,120°)$
7. $(4,-180°)$	**8.** $(5,-90°)$	**9.** $(-6,225°)$
10. $(-2,-180°)$	**11.** $(-3,-225°)$	**12.** $(-4,270°)$

10-9   Graphs of polar coordinate equations

In this section we shall consider the problem of plotting the graphs of equations expressed in polar coordinates.

Definition 10-5   *The graph of an equation in polar coordinates consists of the set of all points whose coordinates satisfy the equation.*

In Sec. 10-8 we pointed out that the r coordinate has the same absolute value at all points on a circle with center at the pole. Hence the graph of the equation $r = a$, a constant, is the circle of radius a with the center at the pole. The graph of an equation in which the θ coordinate is an angle of constant measure is also simply obtained. Thus the graph of $\theta = 45°$ is the line through the pole making an angle of $45°$ with the polar axis. This is true because the coordinates of any point of the line may be expressed with $\theta = 45°$ by selecting the proper value for r. As examples, the coordinates of the point on the terminal side two units from the origin may be expressed by $(2,45°)$ and the point two units from the origin on the extension of the terminal side, by $(-2,45°)$.

EXAMPLE 1   Draw the graph of the equation $r \cos \theta = a$, $a > 0$.

Solution   Noting that $\cos 0° = 1$, we see that the radius vector r is equal to a when θ is equal to $0°$. Hence the point $P(a,0°)$ is on the graph of the given equation. Through P we draw a line perpendicular to the polar axis (Fig. 10-14). This line, as we will show, is the desired graph. Any point Q of the line has (r,θ) as its coordinates. From the right triangle OPQ, we see that $\cos \theta = a/r$, and therefore $r \cos \theta = a$. Now we conclude that the point Q, and consequently all points of the line PQ, belong to the graph of the given equation. It would be easy to show that any point off the line is not a part of the graph. Accordingly, the line PQ is the complete graph of the equation $r \cos \theta = a$.

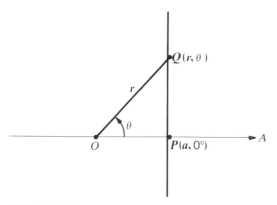

FIGURE 10-14

How would the graph be altered if a were negative? We suggest that the student draw the graph of $r \cos \theta = a$, $a < 0$.

EXAMPLE 2   Draw the graph of the equation $r = a \cos \theta$, $a > 0$.

Solution   We could construct the graph by first preparing a table of corresponding values of r and θ. As we shall show, however, the graph is a circle of which the pole and the point $(a,0°)$ are ends of a diameter. The circle then can be drawn without plotting other points. If the terminal side of θ is in the first quadrant, the corresponding point of the graph is in the first quadrant, as represented in Fig. 10-15. Applying the law of cosines to the triangle in the figure and using the equation $r = a \cos \theta$, we have

$$s^2 = r^2 + \frac{a^2}{4} - ar \cos \theta$$

$$= r^2 + \frac{a^2}{4} - r^2$$

$$s = \frac{a}{2}$$

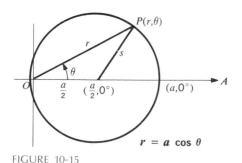

FIGURE 10-15

Clearly, then, each angle θ in the first quadrant yields a point of the graph at a distance $\frac{1}{2}a$ from the point $(\frac{1}{2}a,0°)$. Corresponding to each θ there is an angle $(180° + \theta)$ in the third quadrant which yields the same point as θ does. This is true because $\cos(180° + \theta) = -\cos\theta$, which makes r negative. We conclude therefore that all points of the graph for angles in the first and third quadrants lie on a semicircle in the first quadrant. Similarly, all points of the graph corresponding to angles in the second and fourth quadrants are on a semicircle in the fourth quadrant.

It is quite easy to prove that all points of the circle of radius $\frac{1}{2}a$ and center at $(\frac{1}{2}a,0°)$ belong to the graph. We suggest that the student establish this fact. Since all points of the graph are on the circle and all points of the circle belong to the graph, the circle must be the complete graph of the given equation.

EXAMPLE 3   Draw the graph of the equation $r = 2 + \cos\theta$.

Solution   We assign certain values to θ in the interval $0°$ to $360°$ and prepare the following table.

θ	0°	30°	45°	60°	90°	120°	150°	180°
r	3	2.9	2.7	2.5	2	1.5	1.1	1

θ	210°	225°	240°	270°	300°	315°	330°	360°
r	1.1	1.3	1.5	2	2.5	2.7	2.9	3

By drawing a smooth curve through the points determined from the table, we obtain the graph of Fig. 10-16.

Many polar coordinate equations are sufficiently simple so that only a very few points need be plotted in order to draw their graphs. We illustrate this situation in an example.

EXAMPLE 4   Draw the graph of the equation $r = 1 + 2\cos\theta$.

Solution   Selecting values for θ such that the corresponding values of r can be readily obtained, we prepare a table of values.

θ	0°	60°	90°	120°	180°	240°	270°	330°	360°
r	3	2	1	0	−1	0	1	2	3

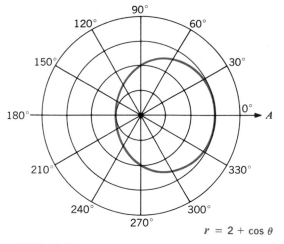

$$r = 2 + \cos \theta$$

FIGURE 10-16

From the points determined by these tabulated values and a knowledge of the way $\cos \theta$ varies as θ increases from each angle to the next one, a satisfactory graph can be drawn (Fig. 10-17). We remark that angles 120° and 240° yield zero values for r and consequently are most helpful in drawing the inner loop of the graph.

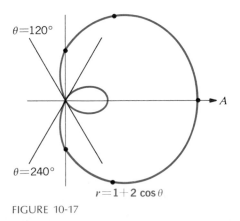

$$r = 1 + 2 \cos \theta$$

FIGURE 10-17

Exercise 10-6

Describe the graph of each equation 1 through 6.

1. $r = -4$ **2.** $r = -3$ **3.** $r = 0$
4. $\theta = 45°$ **5.** $\theta = -90°$ **6.** $\theta = 180°$

The graph of each equation 7 through 12 is a straight line. Draw the line.

7. $r \cos \theta = 3$ **8.** $r \cos \theta = -3$ **9.** $r = 4 \sec \theta$
10. $r \sin \theta = 5$ **11.** $r \sin \theta = -4$ **12.** $r = 3 \sec \theta$

The graph of each of the following equations 13 through 18 is a circle. Draw the circle.

13. $r = 4 \cos \theta$ **14.** $r = -3 \cos \theta$ **15.** $r = \cos \theta$
16. $r = 3 \sin \theta$ **17.** $r = -3 \sin \theta$ **18.** $r = \sin \theta$

Make a table of corresponding values of r and θ and construct the graph of each equation.

19. $r = 3 - \cos \theta$ **20.** $r = 4 + \sin \theta$
21. $r = 2 - \sin \theta$ **22.** $2 - \cos \theta$
23. $r = 2 + \cos \theta$ **24.** $r = 2 + \sin \theta$
25. $r = 1 + 2 \sin \theta$ **26.** $r = 1 - 2 \cos \theta$
27. $r = \tan \theta, 0° \leq \theta < 90°$ **28.** $r = \frac{1}{2} \cot \theta, 90° \geq \theta > 0°$

10-10 Relations between rectangular coordinates and polar coordinates

In many problems or investigations it is advantageous to shift from one coordinate system to another. We shall now show how to transform polar coordinates into rectangular coordinates and rectangular coordinates into polar coordinates.

In Fig. 10-18 the two systems of coordinates are placed so that the origins coincide and the polar axis is along the positive x axis. We denote the rectangular coordinates of the point P by (x,y) and the polar coordinates by (r,θ). From the right triangle OMP, we have at once

$$\cos \theta = \frac{x}{r} \quad \text{and} \quad \sin \theta = \frac{y}{r}$$

or

$$x = r \cos \theta \qquad\qquad (3)$$
$$y = r \sin \theta \qquad\qquad (4)$$

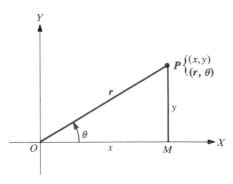

FIGURE 10-18

Referring to the figure again, we write

$$r^2 = x^2 + y^2 \qquad \text{and} \qquad \tan \theta = \frac{y}{x}$$

These equations yield

$$r = \pm \sqrt{x^2 + y^2} \tag{5}$$

$$\theta = \tan^{-1} \frac{y}{x} \tag{6}$$

The preceding formulas enable us to transform the coordinates of a point from one system to the other. Further, they enable us to transform an equation in polar coordinates to one in rectangular coordinates, and vice versa.

EXAMPLE 1   Find the rectangular coordinates of the point defined by the polar coordinates $(6,150°)$.

Solution   From formulas (3) and (4), we find

$$x = r \cos \theta = 6 \cos 150° = -3\sqrt{3}$$
$$y = r \sin \theta = 6 \sin 150° = 3$$

EXAMPLE 2   Find polar coordinates of the point determined by the rectangular coordinates $(-2, -2\sqrt{3})$.

Solution   Formulas (5) and (6) yield

$$r = \sqrt{x^2 + y^2} = 4 \qquad \text{and} \qquad \theta = \tan^{-1} \sqrt{3}$$

Since the given point is in the third quadrant, we choose $\theta = 240°$. Hence $(4,240°)$ is a polar representation of the given point.

EXAMPLE 3   Express the equation $2x + y = 4$ in terms of polar coordinates.

Solution   Substituting for x and y, we obtain

$$2r \cos \theta + r \sin \theta = 4$$

or

$$r = \frac{4}{2 \cos \theta + \sin \theta}$$

EXAMPLE 4   Transform the equation $r = 5 \cos \theta$ to rectangular coordinates.

Solution   Using $\cos \theta = x/r$ and $r^2 = x^2 + y^2$, we have

$$r = \frac{5x}{r} \qquad r^2 = 5x \qquad x^2 + y^2 = 5x$$

Exercise 10-7

Find the rectangular coordinates of the points which have the following polar coordinates.

1. $(3, 90°)$ 2. $(\sqrt{2}, 45°)$ 3. $(4, 0°)$
4. $(0, 40°)$ 5. $(-2, 270°)$ 6. $(1, -30°)$

Find nonnegative polar coordinates of the points which have the following rectangular coordinates.

7. $(0, 3)$ 8. $(5, 0)$ 9. $(-2, 0)$
10. $(0, 0)$ 11. $(\sqrt{2}, -\sqrt{2})$ 12. $(-6, 6\sqrt{3})$
13. $(2, -2\sqrt{3})$ 14. $(-1, -\sqrt{3})$ 15. $(4, 3)$

Transform the following equations to the corresponding polar coordinate equations.

16. $x = 5$ 17. $y = -4$ 18. $y = -x$
19. $2x - 3y = 6$ 20. $x^2 + y^2 = 4$ 21. $3xy = 2$

Transform the following equations to the corresponding rectangular coordinate equations.

22. $r = 5$ 23. $\theta = 135°$ 24. $r \cos \theta = 4$
25. $r \sin \theta = -6$ 26. $r = 2 \sec \theta$ 27. $r = 2 \cos \theta$

Chapter 10   Review exercise

Solve each equation for x and y.

1. $y - 2x + 3i = (2x + y)i$ 2. $7y + (y - 3)i = -3x + 9i$

Perform the indicated operations, leaving each result in the form $a + bi$.

3. $(3 - 2i) + (4 + i) - (2 + i)$ **4.** $(3 + 8i) - 2i - (2 + 7i)$

5. $(3 + 2i)(2 - 4i)$ **6.** $(5 + 4i)(4 - 5i)$

7. $\dfrac{12 - 6i}{2i}$ **8.** $\dfrac{3 + i}{2 - 3i}$ **9.** $\dfrac{7 - i}{4 - 3i}$

Perform the indicated operations. Leave each result as an ordered pair.

10. $(4, -3) + (2, 3)$ **11.** $(3, 4) \cdot (2, 0)$

12. $\dfrac{(3, 2)}{(2, 1)}$ **13.** $\dfrac{(1, 3)}{(4, 2)}$ **14.** $\dfrac{(0, 1)}{(2, 0)}$

Write the corresponding trigonometric form of each complex number.

15. $\sqrt{3} + i$ **16.** $2 + 2i\sqrt{3}$ **17.** $\sqrt{2} - i\sqrt{2}$

Change each number to the algebraic form.

18. $2(\cos 60° + i \sin 60°)$ **19.** $\sqrt{2}(\cos 45° + i \sin 45°)$

Perform the indicated operations. Leave each result in the form $a + bi$.

20. $3(\cos 35° + i \sin 35°) \cdot 2(\cos 25° + i \sin 25°)$

21. $\dfrac{12(\cos 46° + i \sin 46°)}{3(\cos 16° + i \sin 16°)}$

22. Find the cube roots of one. Leave the results in algebraic form.

Plot the point corresponding to each of the following pairs of coordinates. Write three other pairs of polar coordinates of the point, restricting θ so that $-360° < \theta \leq 360°$.

23. $(4, 60°)$ **24.** $(5, 30°)$ **25.** $(2, 0°)$

Describe the graph of each equation.

26. $r = 5$ **27.** $r = -3$ **28.** $\theta = 240°$ **29.** $\theta = -30°$

Draw the graph of each equation.

30. $r \cos \theta = 3$ **31.** $r = 3 \cos \theta$ **32.** $r = 2 \sin \theta$

33. $r = 2 + \cos \theta$ **34.** $r = 4 - \sin \theta$

appendix

A-1 The law of tangents

In Chap. 8 we used the law of sines and the law of cosines in solving the oblique triangle for the various cases of known parts. The law of sines, as we saw, is in suitable form for the application of logarithms. But the law of cosines is not well suited for logarithmic computations. We now derive alternate laws which are readily adaptable to the use of logarithms.

Theorem A-1 *In any triangle the difference of two sides divided by their sum is equal to the tangent of half the difference of the opposite angles divided by the tangent of half their sum. That is,*

$$\frac{a - b}{a + b} = \frac{\tan \frac{1}{2}(A - B)}{\tan \frac{1}{2}(A + B)}$$

$$\frac{b - c}{b + c} = \frac{\tan \frac{1}{2}(B - C)}{\tan \frac{1}{2}(B + C)}$$

$$\frac{c - a}{c + a} = \frac{\tan \frac{1}{2}(C - A)}{\tan \frac{1}{2}(C + A)}$$

Proof We begin by writing the law of sines in the form

$$\frac{a}{b} = \frac{\sin A}{\sin B}$$

This equation may be expressed in two different forms by subtracting 1 from both members and adding 1 to both members. Thus we have

$$\frac{a}{b} - 1 = \frac{\sin A}{\sin B} - 1 \qquad \text{and} \qquad \frac{a}{b} + 1 = \frac{\sin A}{\sin B} + 1$$

Hence, combining, we obtain

$$\frac{a - b}{b} = \frac{\sin A - \sin B}{\sin B} \quad \text{and} \quad \frac{a + b}{b} = \frac{\sin A + \sin B}{\sin B}$$

These two equations, by division, yield

$$\frac{a - b}{a + b} = \frac{\sin A - \sin B}{\sin A + \sin B}$$

Then applying formulas (39) and (40) of Sec. 4-10, we have

$$\frac{a - b}{a + b} = \frac{2 \cos \frac{1}{2}(A + B) \sin \frac{1}{2}(A - B)}{2 \sin \frac{1}{2}(A + B) \cos \frac{1}{2}(A - B)}$$

$$= \cot \tfrac{1}{2}(A + B) \tan \tfrac{1}{2}(A - B)$$

Hence,

$$\frac{a - b}{a + b} = \frac{\tan \frac{1}{2}(A - B)}{\tan \frac{1}{2}(A + B)}$$

We note that the second of the above forms for the law of tangents may be obtained by the same steps by starting with b and c in the law of sines and the third form by starting with sides c and a.

The law of tangents is applicable to case III where two sides and the included angle are known. Suppose we indicate the two known sides by b and c and the included angle by A. With A given, half the sum of B and C can be readily obtained. Then the equation

$$\frac{b - c}{b + c} = \frac{\tan \frac{1}{2}(B - C)}{\tan \frac{1}{2}(B + C)}$$

contains $\tan \frac{1}{2}(B - C)$ as an unknown and consequently may be solved for $\frac{1}{2}(B - C)$. With $\frac{1}{2}(B + C)$ and $\frac{1}{2}(B - C)$ known, B and C can be found individually. Finally, side a is obtainable by the law of sines.

EXAMPLE   Solve the triangle ABC if $A = 46° \, 20'$, $b = 173$, and $c = 283$.

Solution   Since $c > b$, it is convenient to use the law of tangents in the
form

$$\frac{c - b}{c + b} = \frac{\tan \frac{1}{2}(C - B)}{\tan \frac{1}{2}(C + B)}$$

The given values yield $\frac{1}{2}(C + B) = \frac{1}{2}(180° - A) = 66° \, 50'$, $c - b = 110$, and $c + b = 456$. Then substituting in the chosen form of the law of tangents gives

$$\tan \tfrac{1}{2}(C - B) = \frac{110 \tan 66° \; 50'}{456}$$

and

$$\log \tan \tfrac{1}{2}(C - B) = \log 110 + \log \tan 66° \; 50' - \log 456$$

Computing, we find

$$
\begin{aligned}
\log 110 &= 2.0414 \\
\log \tan 66° \; 50' &= 10.3686 - 10 \\
\hline
&12.4100 - 10 \\
\log 456 &= 2.6590 \\
\hline
\log \tan \tfrac{1}{2}(C - B) &= 9.7510 - 10 \\
\tfrac{1}{2}(C - B) &= 29° \; 24'
\end{aligned}
$$

The sum and the difference of the corresponding members of the equations $\tfrac{1}{2}(C + B) = 66° \; 50'$ and $\tfrac{1}{2}(C - B) = 29° \; 24'$ yields $C = 96° \; 14'$ and $B = 37° \; 26'$. Then side a, obtainable by the law of sines, is 205.9. Rounding off, we give the final results as $B = 37° \; 30'$, $C = 96° \; 10'$, and $a = 206$.

Exercise A-1

Solve by the law of tangents and the law of sines.

1. $a = 18$, $b = 32$, $C = 59°$
2. $a = 74$, $b = 48$, $C = 27°$
3. $b = 81$, $c = 67$, $A = 133°$
4. $b = 44$, $c = 36$, $A = 161°$
5. $a = 63$, $c = 42$, $B = 46°$
6. $a = 90$, $c = 71$, $B = 75°$
7. $b = 90.4$, $c = 60.4$, $A = 91° \; 40'$
8. $a = 415$, $b = 715$, $C = 89° \; 10'$
9. $a = 124$, $c = 150$, $B = 38° \; 30'$
10. $b = 57.0$, $c = 65.1$, $A = 143° \; 20'$
11. $b = 7.071$, $c = 4.032$, $A = 84° \; 53'$
12. $a = 671.2$, $b = 981.7$, $C = 107° \; 30'$
13. $a = 0.3408$, $b = 0.2003$, $C = 118° \; 34'$
14. $b = 1842$, $c = 2450$, $A = 32° \; 35'$.

A-2 The half-angle formulas

We shall now derive formulas by which the angles of a triangle may be computed from the lengths of the sides.

Theorem A-2   *If r is the radius of the inscribed circle of a triangle and s is half the sum of the lengths of the sides, then*

$$\tan \tfrac{1}{2}A = \frac{r}{s-a} \qquad \tan \tfrac{1}{2}B = \frac{r}{s-b}$$

$$\tan \tfrac{1}{2}C = \frac{r}{s-c}$$

Proof   The bisectors of angle A and angle B (Fig. A-1) meet at O. The point O, being on the bisector of A, is equally distant from the sides b and c. Likewise O is equally distant from the sides a and c. Hence O is the same distance from the three sides, and therefore a circle can be inscribed in the triangle. The center, of course, is at O and the radius r is the perpendicular distance to the sides.

We see readily from the figure that

$$\tan \tfrac{1}{2}A = \frac{r}{AF} \qquad \tan \tfrac{1}{2}B = \frac{r}{BD} \qquad \tan \tfrac{1}{2}C = \frac{r}{CE}$$

To complete the derivation, we need to express r, AF, BD, and CE in terms of the sides a, b, and c. The area of triangle ABC is equal to the sum of the areas of triangles BOC, COA, and AOB. Thus,

$$S = \tfrac{1}{2}ar + \tfrac{1}{2}br + \tfrac{1}{2}cr = \tfrac{1}{2}r(a+b+c) = rs$$

where $a + b + c = 2s$. Hence the area of a triangle is equal to the radius of the inscribed circle multiplied by one-half the sum of the sides. If we equate rs to the formula for the area of a triangle (Theorem 8-5), we obtain

$$rs = \sqrt{s(s-a)(s-b)s-c)}$$

and

$$r = \sqrt{\frac{(s-a)(s-b)(s-c)}{s}} \tag{1}$$

The triangles AFO and AEO are congruent, and hence $AF = AE$. Similarly, $BD = BF$ and $CD = CE$. Since the sum of these six segments is equal to $a + b + c$, or $2s$, we have

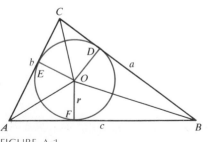

FIGURE A-1

$$2AF + 2BD + 2CD = 2s$$

$$AF = s - (BD + CD) = s - a$$

In a similar way, $BD = s - b$ and $CE = s - c$. Now we may write

$$\tan \tfrac{1}{2}A = \frac{r}{s - a} \qquad \tan \tfrac{1}{2}B = \frac{r}{s - b} \qquad \tan \tfrac{1}{2}C = \frac{r}{s - c}$$

These are the *half-angle formulas*. If the three angles of a triangle are computed by these formulas, then the sum of the angles furnishes a quick check.

EXAMPLE   The sides of a triangle are $a = 14$, $b = 15$, and $c = 20$. Find A, B, and C.

Solution   The quantities s, $s - a$, $s - b$, and $s - c$ need to be computed first. From these values we can find $\log r$ and the logarithms of the denominators in the half-angle formulas. Thus, taking the logarithms of the members of formula (1) above, we have

$$\log r = \tfrac{1}{2}[\log (s - a) + \log (s - b) + \log (s - c) - \log s]$$

We find that $s = \tfrac{1}{2}(a + b + c) = 24.5$. Then we proceed with the following computations.

$$
\begin{array}{ll}
s = 24.5 & \log (s - a) = 1.0212 \\
s - a = 10.5 & (+) \log (s - b) = 0.9777 \\
s - b = 9.5 & (+) \log (s - c) = 0.6532 \\
\cline{2-2}
s - c = 4.5 & \qquad\qquad\qquad 2.6521 \\
& (-) \log s = 1.3892 \\
\cline{2-2}
& \log r^2 = 1.2629 \\
& \log r = 0.6314
\end{array}
$$

We now have the necessary logarithms for applying the half-angle formulas. Thus we get

$$\log \tan \tfrac{1}{2}A = \log r - \log (s - a) = 9.6102 - 10$$

$$\tfrac{1}{2}A = 22° \, 10' \qquad A = 44° \, 20'$$

$$\log \tan \tfrac{1}{2}B = \log r - \log (s - b) = 9.6537 - 10$$

$$\tfrac{1}{2}B = 24° \, 20' \qquad B = 48° \, 40'$$

$$\log \tan \tfrac{1}{2}C = \log r - \log (s - c) = 9.9782 - 10$$

$$\tfrac{1}{2}C = 43° \, 30' \qquad C = 87° \, 00'$$

Rounding off to the nearest degree, we find $A = 44°$, $B = 49°$, and $C = 87°$.

Exercise A-2

Solve the following triangles.

1. $a = 16, b = 25, c = 37$
2. $a = 62, b = 78, c = 86$
3. $a = 13, b = 15, c = 21$
4. $a = 11, b = 13, c = 20$
5. $a = 7.6, b = 6.2, c = 9.8$
6. $a = 0.23, b = 0.31, c = 0.34$
7. $a = 19.0, b = 21.4, c = 27.6$
8. $a = 409, b = 437, c = 532$
9. $a = 520, b = 700, c = 902$
10. $a = 360, b = 210, c = 217$
11. $a = 785.3, b = 564.2, c = 343.1$
12. $a = 40.09, b = 50.17, c = 70.01$

13. Notice in Fig. A-1 that

$$\sin \tfrac{1}{2}A = \frac{r}{\sqrt{r^2 + (s-a)^2}} \qquad \text{and} \qquad \cos \tfrac{1}{2}A = \frac{s-a}{\sqrt{r^2 + (s-a)^2}}$$

Simplify the right members of these equations and obtain

$$\sin \tfrac{1}{2}A = \sqrt{\frac{(s-b)(s-c)}{bc}} \qquad \text{and} \qquad \cos \tfrac{1}{2}A = \sqrt{\frac{s(s-a)}{bc}}$$

Then write, without deriving, the corresponding formulas for these functions of $\tfrac{1}{2}B$ and $\tfrac{1}{2}C$.

table 1 powers and roots

No.	Sq.	Sq. root	Cube	Cube root	No.	Sq.	Sq. root	Cube	Cube root
1	1	1.000	1	1.000	51	2,601	7.141	132,651	3.708
2	4	1.414	8	1.260	52	2,704	7.211	140,608	3.733
3	9	1.732	27	1.442	53	2,809	7.280	148,877	3.756
4	16	2.000	64	1.587	54	2,916	7.348	157,464	3.780
5	25	2.236	125	1.710	55	3,025	7.416	166,375	3.803
6	36	2.449	216	1.817	56	3,136	7.483	175,616	3.826
7	49	2.646	343	1.913	57	3,249	7.550	185,193	3.849
8	64	2.828	512	2.000	58	3,364	7.616	195,112	3.871
9	81	3.000	729	2.080	59	3,481	7.681	205,379	3.893
10	100	3.162	1,000	2.154	60	3,600	7.746	216,000	3.915
11	121	3.317	1,331	2.224	61	3,721	7.810	226,981	3.936
12	144	3.464	1,728	2.289	62	3,844	7.874	238,328	3.958
13	169	3.606	2,197	2.351	63	3,969	7.937	250,047	3.979
14	196	3.742	2,744	2.410	64	4,096	8.000	262,144	4.000
15	225	3.873	3,375	2.466	65	4,225	8.062	274,625	4.021
16	256	4.000	4,096	2.520	66	4,356	8.124	287,496	4.041
17	289	4.123	4,913	2.571	67	4,489	8.185	300,763	4.062
18	324	4.243	5,832	2.621	68	4,624	8.246	314,432	4.082
19	361	4.359	6,859	2.668	69	4,761	8.307	328,509	4.102
20	400	4.472	8,000	2.714	70	4,900	8.367	343,000	4.121
21	441	4.583	9,261	2.759	71	5,041	8.426	357,911	4.141
22	484	4.690	10,648	2.802	72	5,184	8.485	373,248	4.160
23	529	4.796	12,167	2.844	73	5,329	8.544	389,017	4.179
24	576	4.899	13,824	2.884	74	5,476	8.602	405,224	4.198
25	625	5.000	15,625	2.924	75	5,625	8.660	421,875	4.217
26	676	5.099	17,576	2.962	76	5,776	8.718	438,976	4.236
27	729	5.196	19,683	3.000	77	5,929	8.775	456,533	4.254
28	784	5.292	21,952	3.037	78	6,084	8.832	474,552	4.273
29	841	5.385	24,389	3.072	79	6,241	8.888	493,039	4.291
30	900	5.477	27,000	3.107	80	6,400	8.944	512,000	4.309
31	961	5.568	29,791	3.141	81	6,561	9.000	531,441	4.327
32	1,024	5.657	32,768	3.175	82	6,724	9.055	551,368	4.344
33	1,089	5.745	35,937	3.208	83	6,889	9.110	571,787	4.362
34	1,156	5.831	39,304	3.240	84	7,056	9.165	592,704	4.380
35	1,225	5.916	42,875	3.271	85	7,225	9.220	614,125	4.397
36	1,296	6.000	46,656	3.302	86	7,396	9.274	636,056	4.414
37	1,369	6.083	50,653	3.332	87	7,569	9.327	658,503	4.431
38	1,444	6.164	54,872	3.362	88	7,744	9.381	681,472	4.448
39	1,521	6.245	59,319	3.391	89	7,921	9.434	704,969	4.465
40	1,600	6.325	64,000	3.420	90	8,100	9.487	729,000	4.481
41	1,681	6.403	68,921	3.448	91	8,281	9.539	753,571	4.498
42	1,764	6.481	74,088	3.476	92	8,464	9.592	778,688	4.514
43	1,849	6.557	79,507	3.503	93	8,649	9.644	804,357	4.531
44	1,936	6.633	85,184	3.530	94	8,836	9.695	830,584	4.547
45	2,025	6.708	91,125	3.557	95	9,025	9.747	857,375	4.563
46	2,116	6.782	97,336	3.583	96	9,216	9.798	884,736	4.579
47	2,209	6.856	103,823	3.609	97	9,409	9.849	912,673	4.595
48	2,304	6.928	110,592	3.634	98	9,604	9.899	941,192	4.610
49	2,401	7.000	117,649	3.659	99	9,801	9.950	970,299	4.626
50	2,500	7.071	125,000	3.684	100	10,000	10.000	1,000,000	4.642

table 2 natural trigonometric functions

Degrees	Sin	Cos	Tan	Cot	Sec	Csc	
0° 00′	.0000	1.0000	.0000		1.000		**90° 00′**
10	.0029	1.0000	.0029	343.8	1.000	343.8	50
20	.0058	1.0000	.0058	171.9	1.000	171.9	40
30	.0087	1.0000	.0087	114.6	1.000	114.6	30
40	.0116	.9999	.0116	85.94	1.000	85.95	20
50	.0145	.9999	.0145	68.75	1.000	68.76	10
1° 00′	.0175	.9998	.0175	57.29	1.000	57.30	**89° 00′**
10	.0204	.9998	.0204	49.10	1.000	49.11	50
20	.0233	.9997	.0233	42.96	1.000	42.98	40
30	.0262	.9997	.0262	38.19	1.000	38.20	30
40	.0291	.9996	.0291	34.37	1.000	34.38	20
50	.0320	.9995	.0320	31.24	1.001	31.26	10
2° 00′	.0349	.9994	.0349	28.64	1.001	28.65	**88° 00′**
10	.0378	.9993	.0378	26.43	1.001	26.45	50
20	.0407	.9992	.0407	24.54	1.001	24.56	40
30	.0436	.9990	.0437	22.90	1.001	22.93	30
40	.0465	.9989	.0466	21.47	1.001	21.49	20
50	.0494	.9988	.0495	20.21	1.001	20.23	10
3° 00′	.0523	.9986	.0524	19.08	1.001	19.11	**87° 00′**
10	.0552	.9985	.0553	18.07	1.002	18.10	50
20	.0581	.9983	.0582	17.17	1.002	17.20	40
30	.0610	.9981	.0612	16.35	1.002	16.38	30
40	.0640	.9980	.0641	15.60	1.002	15.64	20
50	.0669	.9978	.0670	14.92	1.002	14.96	10
4° 00′	.0698	.9976	.0699	14.30	1.002	14.34	**86° 00′**
10	.0727	.9974	.0729	13.73	1.003	13.76	50
20	.0756	.9971	.0758	13.20	1.003	13.23	40
30	.0785	.9969	.0787	12.71	1.003	12.75	30
40	.0814	.9967	.0816	12.25	1.003	12.29	20
50	.0843	.9964	.0846	11.83	1.004	11.87	10
5° 00′	.0872	.9962	.0875	11.43	1.004	11.47	**85° 00′**
10	.0901	.9959	.0904	11.06	1.004	11.10	50
20	.0929	.9957	.0934	10.71	1.004	10.76	40
30	.0958	.9954	.0963	10.39	1.005	10.43	30
40	.0987	.9951	.0992	10.08	1.005	10.13	20
50	.1016	.9948	.1022	9.788	1.005	9.839	10
6° 00′	.1045	.9945	.1051	9.514	1.006	9.567	**84° 00′**
10	.1074	.9942	.1080	9.255	1.006	9.309	50
20	.1103	.9939	.1110	9.010	1.006	9.065	40
30	.1132	.9936	.1139	8.777	1.006	8.834	30
40	.1161	.9932	.1169	8.556	1.007	8.614	20
50	.1190	.9929	.1198	8.345	1.007	8.405	10
7° 00′	.1219	.9925	.1228	8.144	1.008	8.206	**83° 00′**
10	.1248	.9922	.1257	7.953	1.008	8.016	50
20	.1276	.9918	.1287	7.770	1.008	7.834	40
30	.1305	.9914	.1317	7.596	1.009	7.661	30
40	.1334	.9911	.1346	7.429	1.009	7.496	20
50	.1363	.9907	.1376	7.269	1.009	7.337	10
8° 00′	.1392	.9903	.1405	7.115	1.010	7.185	**82° 00′**
10	.1421	.9899	.1435	6.968	1.010	7.040	50
20	.1449	.9894	.1465	6.827	1.011	6.900	40
30	.1478	.9890	.1495	6.691	1.011	6.765	30
40	.1507	.9886	.1524	6.561	1.012	6.636	20
50	.1536	.9881	.1554	6.435	1.012	6.512	10
9° 00′	.1564	.9877	.1584	6.314	1.012	6.392	**81° 00′**
	Cos	Sin	Cot	Tan	Csc	Sec	Degrees

table 2 natural trigonometric functions (continued)

Degrees	Sin	Cos	Tan	Cot	Sec	Csc	
9° 00'	.1564	.9877	.1584	6.314	1.012	6.392	**81° 00'**
10	.1593	.9872	.1614	6.197	1.013	6.277	50
20	.1622	.9868	.1644	6.084	1.013	6.166	40
30	.1650	.9863	.1673	5.976	1.014	6.059	30
40	.1679	.9858	.1703	5.871	1.014	5.955	20
50	.1708	.9853	.1733	5.769	1.015	5.855	10
10° 00'	.1736	.9848	.1763	5.671	1.015	5.759	**80° 00'**
10	.1765	.9843	.1793	5.576	1.016	5.665	50
20	.1794	.9838	.1823	5.485	1.016	5.575	40
30	.1822	.9833	.1853	5.396	1.017	5.487	30
40	.1851	.9827	.1883	5.309	1.018	5.403	20
50	.1880	.9822	.1914	5.226	1.018	5.320	10
11° 00'	.1908	.9816	.1944	5.145	1.019	5.241	**79° 00'**
10	.1937	.9811	.1974	5.066	1.019	5.164	50
20	.1965	.9805	.2004	4.989	1.020	5.089	40
30	.1994	.9799	.2035	4.915	1.020	5.016	30
40	.2022	.9793	.2065	4.843	1.021	4.945	20
50	.2051	.9787	.2095	4.773	1.022	4.876	10
12° 00'	.2079	.9781	.2126	4.705	1.022	4.810	**78° 00'**
10	.2108	.9775	.2156	4.638	1.023	4.745	50
20	.2136	.9769	.2186	4.574	1.024	4.682	40
30	.2164	.9763	.2217	4.511	1.024	4.620	30
40	.2193	.9757	.2247	4.449	1.025	4.560	20
50	.2221	.9750	.2278	4.390	1.026	4.502	10
13° 00'	.2250	.9744	.2309	4.331	1.026	4.445	**77° 00'**
10	.2278	.9737	.2339	4.275	1.027	4.390	50
20	.2306	.9730	.2370	4.219	1.028	4.336	40
30	.2334	.9724	.2401	4.165	1.028	4.284	30
40	.2363	.9717	.2432	4.113	1.029	4.232	20
50	.2391	.9710	.2462	4.061	1.030	4.182	10
14° 00'	.2419	.9703	.2493	4.011	1.031	4.134	**76° 00'**
10	.2447	.9696	.2524	3.962	1.031	4.086	50
20	.2476	.9689	.2555	3.914	1.032	4.039	40
30	.2504	.9681	.2586	3.867	1.033	3.994	30
40	.2532	.9674	.2617	3.821	1.034	3.950	20
50	.2560	.9667	.2648	3.776	1.034	3.906	10
15° 00'	.2588	.9659	.2679	3.732	1.035	3.864	**75° 00'**
10	.2616	.9652	.2711	3.689	1.036	3.822	50
20	.2644	.9644	.2742	3.647	1.037	3.782	40
30	.2672	.9636	.2773	3.606	1.038	3.742	30
40	.2700	.9628	.2805	3.566	1.039	3.703	20
50	.2728	.9621	.2836	3.526	1.039	3.665	10
16° 00'	.2756	.9613	.2867	3.487	1.040	3.628	**74° 00'**
10	.2784	.9605	.2899	3.450	1.041	3.592	50
20	.2812	.9596	.2931	3.412	1.042	3.556	40
30	.2840	.9588	.2962	3.376	1.043	3.521	30
40	.2868	.9580	.2994	3.340	1.044	3.487	20
50	.2896	.9572	.3026	3.305	1.045	3.453	10
17° 00'	.2924	.9563	.3057	3.271	1.046	3.420	**73° 00'**
10	.2952	.9555	.3089	3.237	1.047	3.388	50
20	.2979	.9546	.3121	3.204	1.048	3.356	40
30	.3007	.9537	.3153	3.172	1.049	3.326	30
40	.3035	.9528	.3185	3.140	1.049	3.295	20
50	.3062	.9520	.3217	3.108	1.050	3.265	10
18° 00'	.3090	.9511	.3249	3.078	1.051	3.236	**72° 00'**
	Cos	Sin	Cot	Tan	Csc	Sec	Degrees

table 2　**natural trigonometric functions** (continued)

Degrees	Sin	Cos	Tan	Cot	Sec	Csc	
18° 00'	.3090	.9511	.3249	3.078	1.051	3.236	**72° 00'**
10	.3118	.9502	.3281	3.047	1.052	3.207	50
20	.3145	.9492	.3314	3.018	1.053	3.179	40
30	.3173	.9483	.3346	2.989	1.054	3.152	30
40	.3201	.9474	.3378	2.960	1.056	3.124	20
50	.3228	.9465	.3411	2.932	1.057	3.098	10
19° 00'	.3256	.9455	.3443	2.904	1.058	3.072	**71° 00'**
10	.3283	.9446	.3476	2.877	1.059	3.046	50
20	.3311	.9436	.3508	2.850	1.060	3.021	40
30	.3338	.9426	.3541	2.824	1.061	2.996	30
40	.3365	.9417	.3574	2.798	1.062	2.971	20
50	.3393	.9407	.3607	2.773	1.063	2.947	10
20° 00'	.3420	.9397	.3640	2.747	1.064	2.924	**70° 00'**
10	.3448	.9387	.3673	2.723	1.065	2.901	50
20	.3475	.9377	.3706	2.699	1.066	2.878	40
30	.3502	.9367	.3739	2.675	1.068	2.855	30
40	.3529	.9356	.3772	2.651	1.069	2.833	20
50	.3557	.9346	.3805	2.628	1.070	2.812	10
21° 00'	.3584	.9336	.3839	2.605	1.071	2.790	**69° 00'**
10	.3611	.9325	.3872	2.583	1.072	2.769	50
20	.3638	.9315	.3906	2.560	1.074	2.749	40
30	.3665	.9304	.3939	2.539	1.075	2.729	30
40	.3692	.9293	.3973	2.517	1.076	2.709	20
50	.3719	.9283	.4006	2.496	1.077	2.689	10
22° 00'	.3746	.9272	.4040	2.475	1.079	2.669	**68° 00'**
10	.3773	.9261	.4074	2.455	1.080	2.650	50
20	.3800	.9250	.4108	2.434	1.081	2.632	40
30	.3827	.9239	.4142	2.414	1.082	2.613	30
40	.3854	.9228	.4176	2.394	1.084	2.595	20
50	.3881	.9216	.4210	2.375	1.085	2.577	10
23° 00'	.3907	.9205	.4245	2.356	1.086	2.559	**67° 00'**
10	.3934	.9194	.4279	2.337	1.088	2.542	50
20	.3961	.9182	.4314	2.318	1.089	2.525	40
30	.3987	.9171	.4348	2.300	1.090	2.508	30
40	.4014	.9159	.4383	2.282	1.092	2.491	20
50	.4041	.9147	.4417	2.264	1.093	2.475	10
24° 00'	.4067	.9135	.4452	2.246	1.095	2.459	**66° 00'**
10	.4094	.9124	.4487	2.229	1.096	2.443	50
20	.4120	.9112	.4522	2.211	1.097	2.427	40
30	.4147	.9100	.4557	2.194	1.099	2.411	30
40	.4173	.9088	.4592	2.177	1.100	2.396	20
50	.4200	.9075	.4628	2.161	1.102	2.381	10
25° 00'	.4226	.9063	.4663	2.145	1.103	2.366	**65° 00'**
10	.4253	.9051	.4699	2.128	1.105	2.352	50
20	.4279	.9038	.4734	2.112	1.106	2.337	40
30	.4305	.9026	.4770	2.097	1.108	2.323	30
40	.4331	.9013	.4806	2.081	1.109	2.309	20
50	.4358	.9001	.4841	2.066	1.111	2.295	10
26° 00'	.4384	.8988	.4877	2.050	1.113	2.281	**64° 00'**
10	.4410	.8975	.4913	2.035	1.114	2.268	50
20	.4436	.8962	.4950	2.020	1.116	2.254	40
30	.4462	.8949	.4986	2.006	1.117	2.241	30
40	.4488	.8936	.5022	1.991	1.119	2.228	20
50	.4514	.8923	.5059	1.977	1.121	2.215	10
27° 00'	.4540	.8910	.5095	1.963	1.122	2.203	**63° 00'**
	Cos	Sin	Cot	Tan	Csc	Sec	Degrees

table 2 **natural trigonometric functions** (continued)

Degrees	Sin	Cos	Tan	Cot	Sec	Csc	
27° 00′	.4540	.8910	.5095	1.963	1.122	2.203	**63° 00′**
10	.4566	.8897	.5132	1.949	1.124	2.190	50
20	.4592	.8884	.5169	1.935	1.126	2.178	40
30	.4617	.8870	.5206	1.921	1.127	2.166	30
40	.4643	.8857	.5243	1.907	1.129	2.154	20
50	.4669	.8843	5280	1.894	1.131	2.142	10
28° 00′	.4695	.8829	.5317	1.881	1.133	2.130	**62° 00′**
10	.4720	.8816	.5354	1.868	1.134	2.118	50
20	.4746	.8802	.5392	1.855	1.136	2.107	40
30	.4772	.8788	.5430	1.842	1.138	2.096	30
40	.4797	.8774	.5467	1.829	1.140	2.085	20
50	.4823	.8760	.5505	1 816	1.142	2.074	10
29° 00′	.4848	.8746	.5543	1.804	1.143	2.063	**61° 00′**
10	.4874	.8732	.5581	1.792	1.145	2.052	50
20	.4899	.8718	.5619	1.780	1.147	2.041	40
30	.4924	.8704	.5658	1.767	1.149	2.031	30
40	.4950	.8689	.5696	1.756	1.151	2.020	20
50	.4975	.8675	.5735	1.744	1.153	2.010	10
30° 00′	.5000	.8660	.5774	1.732	1.155	2.000	**60° 00′**
10	.5025	.8646	.5812	1.720	1.157	1.990	50
20	.5050	.8631	.5851	1.709	1.159	1.980	40
30	.5075	.8616	.5890	1.698	1.161	1.970	30
40	.5100	.8601	.5930	1.686	1.163	1.961	20
50	.5125	.8587	.5969	1.675	1.165	1.951	10
31° 00′	.5150	.8572	.6009	1.664	1.167	1.942	**59° 00′**
10	.5175	.8557	.6048	1.653	1.169	1.932	50
20	.5200	.8542	.6088	1.643	1.171	1.923	40
30	.5225	.8526	.6128	1.632	1.173	1.914	30
40	.5250	.8511	.6168	1.621	1.175	1.905	20
50	.5275	.8496	.6208	1.611	1.177	1.896	10
32° 00′	.5299	.8480	.6249	1.600	1.179	1.887	**58° 00′**
10	.5324	.8465	.6289	1.590	1.181	1.878	50
20	.5348	.8450	.6330	1.580	1.184	1.870	40
30	.5373	.8434	.6371	1.570	1.186	1.861	30
40	.5398	.8418	.6412	1.560	1.188	1.853	20
50	.5422	.8403	.6453	1.550	1.190	1.844	10
33° 00′	.5446	.8387	.6494	1.540	1.192	1.836	**57° 00′**
10	.5471	.8371	.6536	1.530	1.195	1.828	50
20	.5495	.8355	.6577	1.520	1.197	1.820	40
30	.5519	.8339	.6619	1.511	1.199	1.812	30
40	.5544	.8323	.6661	1.501	1.202	1.804	20
50	.5568	.8307	.6703	1.492	1.204	1.796	10
34° 00′	.5592	.8290	.6745	1.483	1.206	1.788	**56° 00′**
10	.5616	.8274	.6787	1.473	1.209	1.781	50
20	.5640	.8258	.6830	1.464	1.211	1.773	40
30	.5664	.8241	.6873	1.455	1.213	1.766	30
40	.5688	.8225	.6916	1.446	1.216	1.758	20
50	.5712	.8208	.6959	1.437	1.218	1.751	10
35° 00′	.5736	.8192	.7002	1.428	1.221	1.743	**55° 00′**
10	.5760	.8175	.7046	1.419	1.223	1.736	50
20	.5783	.8158	.7089	1.411	1.226	1.729	40
30	.5807	.8141	.7133	1.402	1.228	1.722	30
40	.5831	.8124	.7177	1.393	1.231	1.715	20
50	.5854	.8107	.7221	1.385	1.233	1.708	10
36° 00′	.5878	.8090	.7265	1.376	1.236	1.701	**54° 00′**
	Cos	Sin	Cot	Tan	Csc	Sec	Degrees

table 2   natural trigonometric functions (continued)

Degrees	Sin	Cos	Tan	Cot	Sec	Csc	
36° 00'	.5878	.8090	.7265	1.376	1.236	1.701	**54° 00'**
10	.5901	.8073	.7310	1.368	1.239	1.695	50
20	.5925	.8056	.7355	1.360	1.241	1.688	40
30	.5948	.8039	.7400	1.351	1.244	1.681	30
40	.5972	.8021	.7445	1.343	1.247	1.675	20
50	.5995	.8004	.7490	1.335	1.249	1.668	10
37° 00'	.6018	.7986	.7536	1.327	1.252	1.662	**53° 00'**
10	.6041	.7969	.7581	1.319	1.255	1.655	50
20	.6065	.7951	.7627	1.311	1.258	1.649	40
30	.6088	.7934	.7673	1.303	1.260	1.643	30
40	.6111	.7916	.7720	1.295	1.263	1.636	20
50	.6134	.7898	.7766	1.288	1.266	1.630	10
38° 00'	.6157	.7880	.7813	1.280	1.269	1.624	**52° 00'**
10	.6180	.7862	.7860	1.272	1.272	1.618	50
20	.6202	.7844	.7907	1.265	1.275	1.612	40
30	.6225	.7826	.7954	1.257	1.278	1.606	30
40	.6248	.7808	.8002	1.250	1.281	1.601	20
50	.6271	.7790	.8050	1.242	1.284	1.595	10
39° 00'	.6293	.7771	.8098	1.235	1.287	1.589	**51° 00'**
10	.6316	.7753	.8146	1.228	1.290	1.583	50
20	.6338	.7735	.8195	1.220	1.293	1.578	40
30	.6361	.7716	.8243	1.213	1.296	1.572	30
40	.6383	.7698	.8292	1.206	1.299	1.567	20
50	.6406	.7679	.8342	1.199	1.302	1.561	10
40° 00'	.6428	.7660	.8391	1.192	1.305	1.556	**50° 00'**
10	.6450	.7642	.8441	1.185	1.309	1.550	50
20	.6472	.7623	.8491	1.178	1.312	1.545	40
30	.6494	.7604	.8541	1.171	1.315	1.540	30
40	.6517	.7585	.8591	1.164	1.318	1.535	20
50	.6539	.7566	.8642	1.157	1.322	1.529	10
41° 00'	.6561	.7547	.8693	1.150	1.325	1.524	**49° 00'**
10	.6583	.7528	.8744	1.144	1.328	1.519	50
20	.6604	.7509	.8796	1.137	1.332	1.514	40
30	.6626	.7490	.8847	1.130	1.335	1.509	30
40	.6648	.7470	.8899	1.124	1.339	1.504	20
50	.6670	.7451	.8952	1.117	1.342	1.499	10
42° 00'	.6691	.7431	.9004	1.111	1.346	1.494	**48° 00'**
10	.6713	.7412	.9057	1.104	1.349	1.490	50
20	.6734	.7392	.9110	1.098	1.353	1.485	40
30	.6756	.7373	.9163	1.091	1.356	1.480	30
40	.6777	.7353	.9217	1.085	1.360	1.476	20
50	.6799	.7333	.9271	1.079	1.364	1.471	10
43° 00'	.6820	.7314	.9325	1.072	1.367	1.466	**47° 00'**
10	.6841	.7294	.9380	1.066	1.371	1.462	50
20	.6862	.7274	.9435	1.060	1.375	1.457	40
30	.6884	.7254	.9490	1.054	1.379	1.453	30
40	.6905	.7234	.9545	1.048	1.382	1.448	20
50	.6926	.7214	.9601	1.042	1.386	1.444	10
44° 00'	.6947	.7193	.9657	1.036	1.390	1.440	**46° 00'**
10	.6967	.7173	.9713	1.030	1.394	1.435	50
20	.6988	.7153	.9770	1.024	1.398	1.431	40
30	.7009	.7133	.9827	1.018	1.402	1.427	30
40	.7030	.7112	.9884	1.012	1.406	1.423	20
50	.7050	.7092	.9942	1.006	1.410	1.418	10
45° 00'	.7071	.7071	1.000	1.000	1.414	1.414	**45° 00'**
	Cos	Sin	Cot	Tan	Csc	Sec	Degrees

table 3   values of trigonometric functions (radians)

Radians	Sin	Cos	Tan	Radians	Sin	Cos	Tan
.00	.0000	1.0000	.0000	**.40**	.3894	.9211	.4228
.01	.0100	1.0000	.0100	.41	.3986	.9171	.4346
.02	.0200	.9998	.0200	.42	.4078	.9131	.4466
.03	.0300	.9996	.0300	.43	.4169	.9090	.4586
.04	.0400	.9992	.0400	.44	.4259	.9048	.4708
.05	.0500	.9988	.0500	**.45**	.4350	.9004	.4831
.06	.0600	.9982	.0601	.46	.4439	.8961	.4954
.07	.0699	.9976	.0701	.47	.4529	.8916	.5080
.08	.0799	.9968	.0802	.48	.4618	.8870	.5206
.09	.0899	.9960	.0902	.49	.4706	.8823	.5334
.10	.0998	.9950	.1003	**.50**	.4794	.8776	.5463
.11	.1098	.9940	.1104	.51	.4882	.8727	.5594
.12	.1197	.9928	.1206	.52	.4969	.8678	.5726
.13	.1296	.9916	.1307	.53	.5055	.8628	.5859
.14	.1395	.9902	.1409	.54	.5141	.8577	.5994
.15	.1494	.9888	.1511	**.55**	.5227	.8525	.6131
.16	.1593	.9872	.1614	.56	.5312	.8473	.6269
.17	.1692	.9856	.1717	.57	.5396	.8419	.6410
.18	.1790	.9838	.1820	.58	.5480	.8365	.6552
.19	.1889	.9820	.1923	.59	.5564	.8309	.6696
.20	.1987	.9801	.2027	**.60**	.5646	.8253	.6841
.21	.2085	.9780	.2131	.61	.5729	.8196	.6989
.22	.2182	.9759	.2236	.62	.5810	.8139	.7139
.23	.2280	.9737	.2341	.63	.5891	.8080	.7291
.24	.2377	.9713	.2447	.64	.5972	.8021	.7445
.25	.2474	.9689	.2553	**.65**	.6052	.7961	.7602
.26	.2571	.9664	.2660	.66	.6131	.7900	.7761
.27	.2667	.9638	.2768	.67	.6210	.7838	.7923
.28	.2764	.9611	.2876	.68	.6288	.7776	.8087
.29	.2860	.9582	.2984	.69	.6365	.7712	.8253
.30	.2955	.9553	.3093	**.70**	.6442	.7648	.8423
.31	.3051	.9523	.3203	.71	.6518	.7584	.8595
.32	.3146	.9492	.3314	.72	.6594	.7518	.8771
.33	.3240	.9460	.3425	.73	.6669	.7452	.8949
.34	.3335	.9428	.3537	.74	.6743	.7385	.9131
.35	.3429	.9394	.3650	**.75**	.6816	.7317	.9316
.36	.3523	.9359	.3764	.76	.6889	.7248	.9505
.37	.3616	.9323	.3879	.77	.6961	.7179	.9697
.38	.3709	.9287	.3994	.78	.7033	.7109	.9893
.39	.3802	.9249	.4111	.79	.7104	.7038	1.009

table 3   **values of trigonometric functions (radians)** (continued)

Radians	Sin	Cos	Tan	Radians	Sin	Cos	Tan
.80	.7174	.6967	1.030	**1.20**	.9320	.3624	2.572
.81	.7243	.6895	1.050	1.21	.9356	.3530	2.650
.82	.7311	.6822	1.072	1.22	.9391	.3436	2.733
.83	.7379	.6749	1.093	1.23	.9425	.3342	2.820
.84	.7446	.6675	1.116	1.24	.9458	.3248	2.912
.85	.7513	.6600	1.138	**1.25**	.9490	.3153	3.010
.86	.7578	.6524	1.162	1.26	.9521	.3058	3.113
.87	.7643	.6448	1.185	1.27	.9551	.2963	3.224
.88	.7707	.6372	1.210	1.28	.9580	.2867	3.341
.89	.7771	.6294	1.235	1.29	.9608	.2771	3.467
.90	.7833	.6216	1.260	**1.30**	.9636	.2675	3.602
.91	.7895	.6137	1.286	1.31	.9662	.2579	3.747
.92	.7956	.6058	1.313	1.32	.9687	.2482	3.903
.93	.8016	.5978	1.341	1.33	.9711	.2385	4.072
.94	.8076	.5898	1.369	1.34	.9735	.2288	4.256
.95	.8134	.5817	1.398	**1.35**	.9757	.2190	4.455
.96	.8192	.5735	1.428	1.36	.9779	.2092	4.673
.97	.8249	.5653	1.459	1.37	.9799	.1995	4.913
.98	.8305	.5570	1.491	1.38	.9819	.1896	5.177
.99	.8360	.5487	1.524	1.39	.9837	.1798	5.471
1.00	.8415	.5403	1.557	**1.40**	.9854	.1700	5.798
1.01	.8468	.5319	1.592	1.41	.9871	.1601	6.165
1.02	.8521	.5234	1.628	1.42	.9887	.1502	6.581
1.03	.8573	.5148	1.665	1.43	.9901	.1403	7.055
1.04	.8624	.5062	1.704	1.44	.9915	.1304	7.602
1.05	.8674	.4976	1.743	**1.45**	.9927	.1205	8.238
1.06	.8724	.4889	1.784	1.46	.9939	.1106	8.989
1.07	.8772	.4801	1.827	1.47	.9949	.1006	9.887
1.08	.8820	.4713	1.871	1.48	.9959	.0907	10.98
1.09	.8866	.4625	1.917	1.49	.9967	.0807	12.35
1.10	.8912	.4536	1.965	**1.50**	.9975	.0707	14.10
1.11	.8957	.4447	2.014	1.51	.9982	.0608	16.43
1.12	.9001	.4357	2.066	1.52	.9987	.0508	19.67
1.13	.9044	.4267	2.120	1.53	.9992	.0408	24.50
1.14	.9086	.4176	2.176	1.54	.9995	.0308	32.46
1.15	.9128	.4085	2.234	**1.55**	.9998	.0208	48.08
1.16	.9168	.3993	2.296	1.56	.9999	.0108	92.62
1.17	.9208	.3902	2.360	1.57	1.000	.0008	1256.
1.18	.9246	.3809	2.427				
1.19	.9284	.3717	2.498				

ble 4   logarithms of numbers

N	0	1	2	3	4	5	6	7	8	9
10	0000	0043	0086	0128	0170	0212	0253	0294	0334	0374
11	0414	0453	0492	0531	0569	0607	0645	0682	0719	0755
12	0792	0828	0864	0899	0934	0969	1004	1038	1072	1106
13	1139	1173	1206	1239	1271	1303	1335	1367	1399	1430
14	1461	1492	1523	1553	1584	1614	1644	1673	1703	1732
15	1761	1790	1818	1847	1875	1903	1931	1959	1987	2014
16	2041	2068	2095	2122	2148	2175	2201	2227	2253	2279
17	2304	2330	2355	2380	2405	2430	2455	2480	2504	2529
18	2553	2577	2601	2625	2648	2672	2695	2718	2742	2765
19	2788	2810	2833	2856	2878	2900	2923	2945	2967	2989
20	3010	3032	3054	3075	3096	3118	3139	3160	3181	3201
21	3222	3243	3263	3284	3304	3324	3345	3365	3385	3404
22	3424	3444	3464	3483	3502	3522	3541	3560	3579	3598
23	3617	3636	3655	3674	3692	3711	3729	3747	3766	3784
24	3802	3820	3838	3856	3874	3892	3909	3927	3945	3962
25	3979	3997	4014	4031	4048	4065	4082	4099	4116	4133
26	4150	4166	4183	4200	4216	4232	4249	4265	4281	4298
27	4314	4330	4346	4362	4378	4393	4409	4425	4440	4456
28	4472	4487	4502	4518	4533	4548	4564	4579	4594	4609
29	4624	4639	4654	4669	4683	4698	4713	4728	4742	4757
30	4771	4786	4800	4814	4829	4843	4857	4871	4886	4900
31	4914	4928	4942	4955	4969	4983	4997	5011	5024	5038
32	5051	5065	5079	5092	5105	5119	5132	5145	5159	5172
33	5185	5198	5211	5224	5237	5250	5263	5276	5289	5302
34	5315	5328	5340	5353	5366	5378	5391	5403	5416	5428
35	5441	5453	5465	5478	5490	5502	5514	5527	5539	5551
36	5563	5575	5587	5599	5611	5623	5635	5647	5658	5670
37	5682	5694	5705	5717	5729	5740	5752	5763	5775	5786
38	5798	5809	5821	5832	5843	5855	5866	5877	5888	5899
39	5911	5922	5933	5944	5955	5966	5977	5988	5999	6010
40	6021	6031	6042	6053	6064	6075	6085	6096	6107	6117
41	6128	6138	6149	6160	6170	6180	6191	6201	6212	6222
42	6232	6243	6253	6263	6274	6284	6294	6304	6314	6325
43	6335	6345	6355	6365	6375	6385	6395	6405	6415	6425
44	6435	6444	6454	6464	6474	6484	6493	6503	6513	6522
45	6532	6542	6551	6561	6571	6580	6590	6599	6609	6618
46	6628	6637	6646	6656	6665	6675	6684	6693	6702	6712
47	6721	6730	6739	6749	6758	6767	6776	6785	6794	6803
48	6812	6821	6830	6839	6848	6857	6866	6875	6884	6893
49	6902	6911	6920	6928	6937	6946	6955	6964	6972	6981
50	6990	6998	7007	7016	7024	7033	7042	7050	7059	7067
51	7076	7084	7093	7101	7110	7118	7126	7135	7143	7152
52	7160	7168	7177	7185	7193	7202	7210	7218	7226	7235
52	7243	7251	7259	7267	7275	7284	7292	7300	7308	7316
54	7324	7332	7340	7348	7356	7364	7372	7380	7388	7396
N	0	1	2	3	4	5	6	7	8	9

table 4 logarithms of numbers (continued)

N	0	1	2	3	4	5	6	7	8	9
55	7404	7412	7419	7427	7435	7443	7451	7459	7466	7474
56	7482	7490	7497	7505	7513	7520	7528	7536	7543	7551
57	7559	7566	7574	7582	7589	7597	7604	7612	7619	7627
58	7634	7642	7649	7657	7664	7672	7679	7686	7694	7701
59	7709	7716	7723	7731	7738	7745	7752	7760	7767	7774
60	7782	7789	7796	7803	7810	7818	7825	7832	7839	7846
61	7853	7860	7868	7875	7882	7889	7896	7903	7910	7917
62	7924	7931	7938	7945	7952	7959	7966	7973	7980	7987
63	7993	8000	8007	8014	8021	8028	8035	8041	8048	8055
64	8062	8069	8075	8082	8089	8096	8102	8109	8116	8122
65	8129	8136	8142	8149	8156	8162	8169	8176	8182	8189
66	8195	8202	8209	8215	8222	8228	8235	8241	8248	8254
67	8261	8267	8274	8280	8287	8293	8299	8306	8312	8319
68	8325	8331	8338	8344	8351	8357	8363	8370	8376	8382
69	8388	8395	8401	8407	8414	8420	8426	8432	8439	8445
70	8451	8457	8463	8470	8476	8482	8488	8494	8500	8506
71	8513	8519	8525	8531	8537	8543	8549	8555	8561	8567
72	8573	8579	8585	8591	8597	8603	8609	8615	8621	8627
73	8633	8639	8645	8651	8657	8663	8669	8675	8681	8686
74	8692	8698	8704	8710	8716	8722	8727	8733	8739	8745
75	8751	8756	8762	8768	8774	8779	8785	8791	8797	8802
76	8808	8814	8820	8825	8831	8837	8842	8848	8854	8859
77	8865	8871	8876	8882	8887	8893	8899	8904	8910	8915
78	8921	8927	8932	8938	8943	8949	8954	8960	8965	8971
79	8976	8982	8987	8993	8998	9004	9009	9015	9020	9025
80	9031	9036	9042	9047	9053	9058	9063	9069	9074	9079
81	9085	9090	9096	9101	9106	9112	9117	9122	9128	9133
82	9138	9143	9149	9154	9159	9165	9170	9175	9180	9186
83	9191	9196	9201	9206	9212	9217	9222	9227	9232	9238
84	9243	9248	9253	9258	9263	9269	9274	9279	9284	9289
85	9294	9299	9304	9309	9315	9320	9325	9330	9335	9340
86	9345	9350	9355	9360	9365	9370	9375	9380	9385	9390
87	9395	9400	9405	9410	9415	9420	9425	9430	9435	9440
88	9445	9450	9455	9460	9465	9469	9474	9479	9484	9489
89	9494	9499	9504	9509	9513	9518	9523	9528	9533	9538
90	9542	9547	9552	9557	9562	9566	9571	9576	9581	9586
91	9590	9595	9600	9605	9609	9614	9619	9624	9628	9633
92	9638	9643	9647	9652	9657	9661	9666	9671	9675	9680
93	9685	9689	9694	9699	9703	9708	9713	9717	9722	9727
94	9731	9736	9741	9745	9750	9754	9759	9763	9768	9773
95	9777	9782	9786	9791	9795	9800	9805	9809	9814	9818
96	9823	9827	9832	9836	9841	9845	9850	9854	9859	9863
97	9868	9872	9877	9881	9886	9890	9894	9899	9903	9908
98	9912	9917	9921	9926	9930	9934	9939	9943	9948	9952
99	9956	9961	9965	9969	9974	9978	9983	9987	9991	9996
N	0	1	2	3	4	5	6	7	8	9

table 5   logarithms of trigonometric functions

$\longrightarrow$	L Sin	L Tan	L Cot	L Cos	
0°00′				10.0000	**90°00′**
10′	7.4637	7.4637	12.5363	.0000	89°50′
20′	.7648	.7648	.2352	.0000	40′
30′	7.9408	7.9409	12.0591	.0000	30′
40′	8.0658	8.0658	11.9342	.0000	20′
0°50′	.1627	.1627	.8373	10.0000	10′
1°00′	8.2419	8.2419	11.7581	9.9999	**89°00′**
10′	.3088	.3089	.6911	.9999	88°50′
20′	.3668	.3669	.6331	.9999	40′
30′	.4179	.4181	.5819	.9999	30′
40′	.4637	.4638	.5362	.9998	20′
1°50′	.5050	.5053	.4947	.9998	10′
2°00′	8.5428	8.5431	11.4569	9.9997	**88°00′**
10′	.5776	.5779	.4221	.9997	87°50′
20′	.6097	.6101	.3899	.9996	40′
30′	.6397	.6401	.3599	.9996	30′
40′	.6677	.6682	.3318	.9995	20′
2°50′	.6940	.6945	.3055	.9995	10′
3°00′	8.7188	8.7194	11.2806	9.9994	**87°00′**
10′	.7423	.7429	.2571	.9993	86°50′
20′	.7645	.7652	.2348	.9993	40′
30′	.7857	.7865	.2135	.9992	30′
40′	.8059	.8067	.1933	.9991	20′
3°50′	.8251	.8261	.1739	.9990	10′
4°00′	8.8436	8.8446	11.1554	9.9989	**86°00′**
10′	.8613	.8624	.1376	.9989	85°50′
20′	.8783	.8795	.1205	.9988	40′
30′	.8946	.8960	.1040	.9987	30′
40′	.9104	.9118	.0882	.9986	20′
4°50′	.9256	.9272	.0728	.9985	10′
5°00′	8.9403	8.9420	11.0580	9.9983	**85°00′**
10′	.9545	.9563	.0437	.9982	84°50′
20′	.9682	.9701	.0299	.9981	40′
30′	.9816	.9836	.0164	.9980	30′
40′	8.9945	8.9966	11.0034	.9979	20′
5°50′	9.0070	9.0093	10.9907	.9977	10′
6°00′	9.0192	9.0216	10.9784	9.9976	**84°00′**
	L Cos	L Cot	L Tan	L Sin	$\longleftarrow$

table 5   logarithms of trigonometric functions (continued)

$\longrightarrow$	L Sin	L Tan	L Cot	L Cos	
6°00′	9.0192	9.0216	10.9784	9.9976	**84°00′**
10′	.0311	.0336	.9664	.9975	83°50′
20′	.0426	.0453	.9547	.9973	40′
30′	.0539	.0567	.9433	.9972	30′
40′	.0648	.0678	.9322	.9971	20′
6°50′	.0755	.0786	.9214	.9969	10′
7°00′	9.0859	9.0891	10.9109	9.9968	**83°00′**
10′	.0961	.0995	.9005	.9966	82°50′
20′	.1060	.1096	.8904	.9964	40′
30′	.1157	.1194	.8806	.9963	30′
40′	.1252	.1291	.8709	.9961	20′
7°50′	.1345	.1385	.8615	.9959	10′
8°00′	9.1436	9.1478	10.8522	9.9958	**82°00′**
10′	.1525	.1569	.8431	.9956	81°50′
20′	.1612	.1658	.8342	.9954	40′
30′	.1697	.1745	.8255	.9952	30′
40′	.1781	.1831	.8169	.9950	20′
8°50′	.1863	.1915	.8085	.9948	10′
9°00′	9.1943	9.1997	10.8003	9.9946	**81°00′**
10′	.2022	.2078	.7922	.9944	80°50′
20′	.2100	.2158	.7842	.9942	40′
30′	.2176	.2236	.7764	.9940	30′
40′	.2251	.2313	.7687	.9938	20′
9°50′	.2324	.2389	.7611	.9936	10′
10°00′	9.2397	9.2463	10.7537	9.9934	**80°00′**
10′	.2468	.2536	.7464	.9931	79°50′
20′	.2538	.2609	.7391	.9929	40′
30′	.2606	.2680	.7320	.9927	30′
40′	.2674	.2750	.7250	.9924	20′
10°50′	.2740	.2819	.7181	.9922	10′
11°00′	9.2806	9.2887	10.7113	9.9919	**79°00′**
10′	.2870	.2953	.7047	.9917	78°50′
20′	.2934	.3020	.6980	.9914	40′
30′	.2997	.3085	.6915	.9912	30′
40′	.3058	.3149	.6851	.9909	20′
11°50′	.3119	.3212	.6788	.9907	10′
12°00′	9.3179	9.3275	10.6725	9.9904	**78°00′**
	L Cos	L Cot	L Tan	L Sin	$\longleftarrow$

table 5   logarithms of trigonometric functions (continued)

⟶	L Sin	L Tan	L Cot	L Cos	
12°00′	9.3179	9.3275	10.6725	9.9904	**78°00′**
10′	.3238	.3336	.6664	.9901	77°50′
20′	.3296	.3397	.6603	.9899	40′
30′	.3353	.3458	.6542	.9896	30′
40′	.3410	.3517	.6483	.9893	20′
12°50′	.3466	.3576	.6424	.9890	10′
13°00′	9.3521	9.3634	10.6366	9.9887	**77°00′**
10′	.3575	.3691	.6309	.9884	76°50′
20′	.3629	.3748	.6252	.9881	40′
30′	.3682	.3804	.6196	.9878	30′
40′	.3734	.3859	.6141	.9875	20′
13°50′	.3786	.3914	.6086	.9872	10′
14°00′	9.3837	9.3968	10.6032	9.9869	**76°00′**
10′	.3887	.4021	.5979	.9866	75°50′
20′	.3937	.4074	.5926	.9863	40′
30′	.3986	.4127	.5873	.9859	30′
40′	.4035	.4178	.5822	.9856	20′
14°50′	.4083	.4230	.5770	.9853	10′
15°00′	9.4130	9.4281	10.5719	9.9849	**75°00′**
10′	.4177	.4331	.5669	.9846	74°50′
20′	.4223	.4381	.5619	.9843	40′
30′	.4269	.4430	.5570	.9839	30′
40′	.4314	.4479	.5521	.9836	20′
15°50′	.4359	.4527	.5473	.9832	10′
16°00′	9.4403	9.4575	10.5425	9.9828	**74°00′**
10′	.4447	.4622	.5378	.9825	73°50′
20′	.4491	.4669	.5331	.9821	40′
30′	.4533	.4716	.5284	.9817	30′
40′	.4576	.4762	.5238	.9814	20′
16°50′	.4618	.4808	.5192	.9810	10′
17°00′	9.4659	9.4853	10.5147	9.9806	**73°00′**
10′	.4700	.4898	.5102	.9802	72°50′
20′	.4741	.4943	.5057	.9798	40′
30′	.4781	.4987	.5013	.9794	30′
40′	.4821	.5031	.4969	.9790	20′
17°50′	.4861	.5075	.4925	.9786	10′
18°00′	9.4900	9.5118	10.4882	9.9782	**72°00′**
	L Cos	L Cot	L Tan	L Sin	⟵

table 5 **logarithms of trigonometric functions** (continued)

$\longrightarrow$	L Sin	L Tan	L Cot	L Cos	
18°00'	9.4900	9.5118	10.4882	9.9782	**72°00'**
10'	.4939	.5161	.4839	.9778	71°50'
20'	.4977	.5203	.4797	.9774	40'
30'	.5015	.5245	.4755	.9770	30'
40'	.5052	.5287	.4713	.9765	20'
18°50'	.5090	.5329	.4671	.9761	10'
19°00'	9.5126	9.5370	10.4630	9.9757	**71°00'**
10'	.5163	.5411	.4589	.9752	70°50'
20'	.5199	.5451	.4549	.9748	40'
30'	.5235	.5491	.4509	.9743	30'
40'	.5270	.5531	.4469	.9739	20'
19°50'	.5306	.5571	.4429	.9734	10'
20°00'	9.5341	9.5611	10.4389	9.9730	**70°00'**
10'	.5375	.5650	.4350	.9725	69°50'
20'	.5409	.5689	.4311	.9721	40'
30'	.5443	.5727	.4273	.9716	30'
40'	.5477	.5766	.4234	.9711	20'
20°50'	.5510	.5804	.4196	.9706	10'
21°00'	9.5543	9.5842	10.4158	9.9702	**69°00'**
10'	.5576	.5879	.4121	.9697	68°50'
20'	.5609	.5917	.4083	.9692	40'
30'	.5641	.5954	.4046	.9687	30'
40'	.5673	.5991	.4009	.9682	20'
21°50'	.5704	.6028	.3972	.9677	10'
22°00'	9.5736	9.6064	10.3936	9.9672	**68°00'**
10'	.5767	.6100	.3900	.9667	67°50'
20'	.5798	.6136	.3864	.9661	40'
30'	.5828	.6172	.3828	.9656	30'
40'	.5859	.6208	.3792	.9651	20'
22°50'	.5889	.6243	.3757	.9646	10'
23°00'	9.5919	9.6279	10.3721	9.9640	**67°00'**
10'	.5948	.6314	.3686	.9635	66°50'
20'	.5978	.6348	.3652	.9629	40'
30'	.6007	.6383	.3617	.9624	30'
40'	.6036	.6417	.3583	.9618	20'
23°50'	.6065	.6452	.3548	.9613	10'
24°00'	9.6093	9.6486	10.3514	9.9607	**66°00'**
	L Cos	L Cot	L Tan	L Sin	$\longleftarrow$

table 5 **logarithms of trigonometric functions** (continued)

⟶	L Sin	L Tan	L Cot	L Cos	
24°00′	9.6093	9.6486	10.3514	9.9607	**66°00′**
10′	.6121	.6520	.3480	.9602	65°50′
20′	.6149	.6553	.3447	.9596	40′
30′	.6177	.6587	.3413	.9590	30′
40′	.6205	.6620	.3380	.9584	20′
24°50′	.6232	.6654	.3346	.9579	10′
25°00′	9.6259	9.6687	10.3313	9.9573	**65°00′**
10′	.6286	.6720	.3280	.9567	64°50′
20′	.6313	.6752	.3248	.9561	40′
30′	.6340	.6785	.3215	.9555	30′
40′	.6366	.6817	.3183	.9549	20′
25°50′	.6392	.6850	.3150	.9543	10′
26°00′	9.6418	9.6882	10.3118	9.9537	**64°00′**
10′	.6444	.6914	.3086	.9530	63°50′
20′	.6470	.6946	.3054	.9524	40′
30′	.6495	.6977	.3023	.9518	30′
40′	.6521	.7009	.2991	.9512	20′
26°50′	.6546	.7040	.2960	.9505	10′
27°00′	9.6570	9.7072	10.2928	9.9499	**63°00′**
10′	.6595	.7103	.2897	.9492	62°50′
20′	.6620	.7134	.2866	.9486	40′
30′	.6644	.7165	.2835	.9479	30′
40′	.6668	.7196	.2804	.9473	20′
27°50′	.6692	.7226	.2774	.9466	10′
28°00′	9.6716	9.7257	10.2743	9.9459	**62°00′**
10′	.6740	.7287	.2713	.9453	61°50′
20′	.6763	.7317	.2683	.9446	40′
30′	.6787	.7348	.2652	.9439	30′
40′	.6810	.7378	.2622	.9432	20′
28°50′	.6833	.7408	.2592	.9425	10′
29°00′	9.6856	9.7438	10.2562	9.9418	**61°00′**
10′	.6878	.7467	.2533	.9411	60°50′
20′	.6901	.7497	.2503	.9404	40′
30′	.6923	.7526	.2474	.9397	30′
40′	.6946	.7556	.2444	.9390	20′
29°50′	.6968	.7585	.2415	.9383	10′
30°00′	9.6990	9.7614	10.2386	9.9375	**60°00′**
	L Cos	L Cot	L Tan	L Sin	⟵

table 5 logarithms of trigonometric functions (continued)

	L Sin	L Tan	L Cot	L Cos	
30°00'	9.6990	9.7614	10.2386	9.9375	**60°00'**
10'	.7012	.7644	.2356	.9368	59°50'
20'	.7033	.7673	.2327	9361	40'
30'	.7055	.7701	.2299	.9353	30'
40'	.7076	.7730	.2270	.9346	20'
30°50'	.7097	.7759	.2241	.9338	10'
31°00'	9.7118	9.7788	10.2212	9.9331	**59°00'**
10'	.7139	.7816	.2184	.9323	58°50'
20'	.7160	.7845	.2155	.9315	40'
30'	.7181	.7873	.2127	.9308	30'
40'	.7201	.7902	.2098	.9300	20'
31°50'	.7222	.7930	.2070	.9292	10'
32°00'	9.7242	9.7958	10.2042	9.9284	**58°00'**
10'	.7262	.7986	.2014	.9276	57°50'
20'	.7282	.8014	.1986	.9268	40'
30'	.7302	.8042	.1958	.9260	30'
40'	.7322	.8070	.1930	.9252	20'
32°50'	.7342	.8097	.1903	.9244	10'
33°00'	9.7361	9.8125	10.1875	9.9236	**57°00'**
10'	.7380	.8153	.1847	.9228	56°50'
20'	.7400	.8180	.1820	.9219	40'
30'	.7419	.8208	.1792	.9211	30'
40'	.7438	.8235	.1765	.9203	20'
33°50'	.7457	.8263	.1737	.9194	10'
34°00'	9.7476	9.8290	10.1710	9.9186	**56°00'**
10'	.7494	.8317	.1683	.9177	55°50'
20'	.7513	.8344	.1656	.9169	40'
30'	.7531	.8371	.1629	.9160	30'
40'	.7550	.8398	.1602	.9151	20'
34°50'	.7568	.8425	.1575	.9142	10'
35°00'	9.7586	9.8452	10.1548	9.9134	**55°00'**
10'	.7604	.8479	.1521	.9125	54°50'
20'	.7622	.8506	.1494	.9116	40'
30'	.7640	.8533	.1467	.9107	30'
40'	.7657	.8559	.1441	.9098	20'
35°50'	.7675	.8586	.1414	.9089	10'
36°00'	9.7692	9.8613	10.1387	9.9080	**54°00'**
	L Cos	L Cot	L Tan	L Sin	⟵

table 5   **logarithms of trigonometric functions** (continued)

$\longrightarrow$	L Sin	L Tan	L Cot	L Cos	
36°00′	9.7692	9.8613	10.1387	9.9080	**54°00′**
10′	.7710	.8639	.1361	.9070	53°50′
20′	.7727	.8666	.1334	.9061	40′
30′	.7744	.8692	.1308	.9052	30′
40′	.7761	.8718	.1282	.9042	20′
36°50′	.7778	.8745	.1255	.9033	10′
37°00′	9.7795	9.8771	10.1229	9.9023	**53°00′**
10′	.7811	.8797	.1203	.9014	52°50′
20′	.7828	.8824	.1176	.9004	40′
30′	.7844	.8850	.1150	.8995	30′
40′	.7861	.8876	.1124	.8985	20′
37°50′	.7877	.8902	.1098	.8975	10′
38°00′	9.7893	9.8928	10.1072	9.8965	**52°00′**
10′	.7910	.8954	.1046	.8955	51°50′
20′	.7926	.8980	.1020	.8945	40′
30′	.7941	.9006	.0994	.8935	30′
40′	.7957	.9032	.0968	.8925	20′
38°50′	.7973	.9058	.0942	.8915	10′
39°00′	9.7989	9.9084	10.0916	9.8905	**51°00′**
10′	.8004	.9110	.0890	.8895	50°50′
20′	.8020	.9135	.0865	.8884	40′
30′	.8035	.9161	.0839	.8874	30′
40′	.8050	.9187	.0813	.8864	20′
39°50′	.8066	.9212	.0788	.8853	10′
40°00′	9.8081	9.9238	10.0762	9.8843	**50°00′**
10′	.8096	.9264	.0736	.8832	49°50′
20′	.8111	.9289	.0711	.8821	40′
30′	.8125	.9315	.0685	.8810	30′
40′	.8140	.9341	.0659	.8800	20′
40°50′	.8155	.9366	.0634	.8789	10′
41°00′	9.8169	9.9392	10.0608	9.8778	**49°00′**
10′	.8184	.9417	.0583	.8767	48°50′
20′	.8198	.9443	.0557	.8756	40′
30′	.8213	.9468	.0532	.8745	30′
40′	.8227	.9494	.0506	.8733	20′
41°50′	.8241	.9519	.0481	.8722	10′
42°00′	9.8255	9.9544	10.0456	9.8711	**48°00′**
	L Cos	L Cot	L Tan	L Sin	$\longleftarrow$

table 5 logarithms of trigonometric functions (continued)

⟶	L Sin	L Tan	L Cot	L Cos	
42°00′	9.8255	9.9544	10.0456	9.8711	**48°00′**
10′	.8269	.9570	.0430	.8699	47°50′
20′	.8283	.9595	.0405	.8688	40′
30′	.8297	.9621	.0379	.8676	30′
40′	.8311	.9646	.0354	.8665	20′
42°50′	.8324	.9671	.0329	.8653	10′
43°00′	9.8338	9.9697	10.0303	9.8641	**47°00′**
10′	.8351	.9722	.0278	.8629	46°50′
20′	.8365	.9747	.0253	.8618	40′
30′	.8378	.9772	.0228	.8606	30′
40′	.8391	.9798	.0202	.8594	20′
43°50′	.8405	.9823	.0177	.8582	10′
44°00′	9.8418	9.9848	10.0152	9.8569	**46°00′**
10′	.8431	.9874	.0126	.8557	45°50′
20′	.8444	.9899	.0101	.8545	40′
30′	.8457	.9924	.0076	.8532	30′
40′	.8469	.9949	.0051	.8520	20′
44°50′	.8482	.9975	.0025	.8507	10′
45°00′	9.8495	10.0000	10.0000	9.8495	**45°00′**
	L Cos	L Cot	L Tan	L Sin	⟵

answers to problems whose numbers are not multiples of three

Exercise 1-1   *Page 11*

1. $A = \{$April, June, September, November$\}$
2. $B = \{$Washington, Adams, Jefferson$\}$
4. $\sqrt{41},\ \sqrt{185},\ \sqrt{5}$ 5. $5,\ \sqrt{97},\ \sqrt{26}$
7. $\sqrt{26},\ 17,\ 25$ 8. $2\sqrt{10},\ \sqrt{41},\ \sqrt{37}$
10. ±4 11. ±2 13. $\sqrt{297}$ 14. $\pm\sqrt{39}$
16. 4   17. 6   19. $\sqrt{109}$ 20. 5
22. $AB = \sqrt{80},\ AC = 10,\ BC = 2$
23. $AB = \sqrt{8},\ AC = \sqrt{53},\ BC = \sqrt{41}$
25. $AB = \sqrt{90},\ AC = \sqrt{45},\ BC = \sqrt{153}$
26. $AB = 4,\ AC = 5,\ BC = \sqrt{17}$

Chapter 1   Review exercise   *Page 15*

2. The set satisfies the definition of a function.
4. $x = \pm12$ 5. $y = \pm7$
7. Is a right triangle.   8. Is a right triangle.

Exercise 2-1   *Page 21*

1. $\dfrac{\pi}{9}$ 2. $\dfrac{\pi}{2}$ 4. $\dfrac{5\pi}{3}$ 5. $\dfrac{7\pi}{4}$ 7. $\dfrac{3\pi}{4}$

8. $\dfrac{7\pi}{6}$ 10. $\dfrac{4\pi}{9}$ 11. $\dfrac{5\pi}{4}$ 13. $\dfrac{-7\pi}{18}$ 14. $\dfrac{-7\pi}{60}$

16. $\dfrac{-2\pi}{5}$ 17. 0.1745   19. 0.1483   20. 1.4082

22. 0.5724   23. -1.0854 25. $210°$ 26. $240°$

28. 40°   **29.** −90°   **31.** 54°   **32.** 6°
34. 37½°   **35.** −24°   **37.** −114° 35.5′   **38.** −171° 53.3′
40. −3° 26.3′   **41.** 22° 55.1′   **43.** −12° 36.3′   **44.** 292° 12.6′
46. 11.4 in., 25.5 in., 45.8 in.
47. ½ rad, 1$\frac{3}{7}$ rad, 1$\frac{2}{3}$ rad
49. 4189 miles   **50.** 11,868 miles
52. $\frac{1}{6}$ rad, $\frac{2}{9}$ rad, $\frac{1}{3}$ rad
53. 1.88 in., 6.28 in., 9.42 in.

Exercise 2-2   *Page 23*

5. II, IV, IV, I   **7.** II, III, II, I   **8.** III, IV, III, I
10. 195°, −525°   **11.** 130°, 490°
13. 35°, 395°   **14.** −10°, −370°
16. π, −π **17.** 270°, −90°
19. 225°, −495°   **20.** 330°, −390°
22. 140°, −220°   **23.** 650°, −70°

Exercise 2-3   *Page 30*

Values given in the order sine, cosine, tangent, cotangent, secant, cosecant.

1. $\dfrac{1}{2}, \dfrac{-\sqrt{3}}{2}, \dfrac{-1}{\sqrt{3}}, -\sqrt{3}, \dfrac{-2}{\sqrt{3}}, 2$

2. $\dfrac{-\sqrt{3}}{3}, \dfrac{-\sqrt{6}}{3}, \dfrac{\sqrt{3}}{\sqrt{6}}, \dfrac{\sqrt{6}}{\sqrt{3}}, \dfrac{-3}{\sqrt{6}}, \dfrac{-3}{\sqrt{3}}$

4. $\frac{3}{5}, -\frac{4}{5}, -\frac{3}{4}, -\frac{4}{3}, -\frac{5}{4}, \frac{5}{3}$

5. $-\frac{12}{13}, \frac{5}{13}, -\frac{12}{5}, -\frac{5}{12}, \frac{13}{5}, -\frac{13}{12}$

7. $-\dfrac{5}{\sqrt{41}}, \dfrac{4}{\sqrt{41}}, -\dfrac{5}{4}, -\dfrac{4}{5}, \dfrac{\sqrt{41}}{4}, -\dfrac{\sqrt{41}}{5}$

8. $\dfrac{\sqrt{2}}{3}, \dfrac{\sqrt{7}}{3}, \dfrac{\sqrt{2}}{\sqrt{7}}, \dfrac{\sqrt{7}}{\sqrt{2}}, \dfrac{3}{\sqrt{7}}, \dfrac{3}{\sqrt{2}}$

10. $-\dfrac{7}{\sqrt{74}}, -\dfrac{5}{\sqrt{74}}, \dfrac{7}{5}, \dfrac{5}{7}, -\dfrac{\sqrt{74}}{5}, -\dfrac{\sqrt{74}}{7}$

11. $-\dfrac{5}{\sqrt{34}}, -\dfrac{3}{\sqrt{34}}, \dfrac{5}{3}, \dfrac{3}{5}, -\dfrac{\sqrt{34}}{3}, -\dfrac{\sqrt{34}}{5}$

13. I, II   **14.** II, III   **16.** I, III
17. IV   **19.** IV   **20.** I

22. Tangent and cotangent positive, others negative.

23. Tangent and cotangent positive, others negative.

25. Sine and cosecant positive, others negative.

26. Cosine and secant positive, others negative.

28. Cosine and secant negative, sine and tangent are 0, no cotangent or cosecant.

29. Sine and cosecant positive, others negative.

31. All positive.

32. Cosine and secant positive, others negative.

In problems 34 through 38 the values are for sine, cosine, and tangent in that order.

34. $\frac{5}{13}, -\frac{12}{13}, -\frac{5}{12}$ **35.** $-\frac{15}{17}, -\frac{8}{17}, \frac{15}{8}$

37. $-\dfrac{4}{\sqrt{17}}, \dfrac{1}{\sqrt{17}}, -4$ **38.** $\dfrac{1}{\sqrt{10}}, \dfrac{3}{\sqrt{10}}, \dfrac{1}{3}$

40. $\sin \theta = 0, \tan \theta = 0, \sec \theta = 1$

41. $\sin \theta = 1, \cos \theta = 0, \cot \theta = 0$

43. $\sin 6\pi = 0, \cos 6\pi = 1, \tan 6\pi = 0, \sec 6\pi = 1$

44. $\sin 5\pi = 0, \cos 5\pi = -1, \tan 5\pi = 0, \sec 5\pi = -1$

46. $\sin \dfrac{3\pi}{2} = -1, \cos \dfrac{3\pi}{2} = 0, \cot \dfrac{3\pi}{2} = 0, \csc \dfrac{3\pi}{2} = -1$

The values are given in the order sine, cosine, and tangent.

47. $\frac{40}{41}, \frac{9}{41}, \frac{40}{9}; -\frac{40}{41}, \frac{9}{41}, -\frac{40}{9}$

49. $\frac{7}{25}, -\frac{24}{25}, -\frac{7}{24}; -\frac{7}{25}, \frac{24}{25}, -\frac{7}{24}$

50. $\frac{20}{29}, \frac{21}{29}, \frac{20}{21}; -\frac{20}{29}, \frac{21}{29}, -\frac{20}{21}$

52. $\dfrac{1}{2}, \dfrac{\sqrt{3}}{2}, \dfrac{1}{\sqrt{3}}; \dfrac{1}{2}, -\dfrac{\sqrt{3}}{2}, -\dfrac{1}{\sqrt{3}}$

53. $-\dfrac{2}{3}, -\dfrac{\sqrt{5}}{3}, \dfrac{2}{\sqrt{5}}; -\dfrac{2}{3}, \dfrac{\sqrt{5}}{3}, -\dfrac{2}{\sqrt{5}}$

55. $\dfrac{5}{\sqrt{34}}, \dfrac{3}{\sqrt{34}}, \dfrac{5}{3}; -\dfrac{5}{\sqrt{34}}, -\dfrac{3}{\sqrt{34}}, \dfrac{5}{3}$

Chapter 2   Review exercise   *Page 33*

7. $\dfrac{11\pi}{6}$ **8.** $\dfrac{\pi}{2}$ **10.** 0.85   **11.** 0.178

13. -1.108 **14.** 229° 11.0′   **16.** 45° 50.2′

17. 88° 14.1′   **20.** $\frac{4}{5}, \frac{3}{5}, \frac{4}{3}, \frac{3}{4}, \frac{5}{3}, \frac{5}{4}$

22. $\dfrac{\sqrt{11}}{4}, -\dfrac{\sqrt{5}}{4}, -\dfrac{\sqrt{11}}{\sqrt{5}}, -\dfrac{\sqrt{5}}{\sqrt{11}}, -\dfrac{4}{\sqrt{5}}, \dfrac{4}{\sqrt{11}}$

23. $-\dfrac{\sqrt{6}}{3}, -\dfrac{\sqrt{3}}{3}, \dfrac{\sqrt{6}}{\sqrt{3}}, \dfrac{\sqrt{3}}{\sqrt{6}}, -\dfrac{3}{\sqrt{3}}, -\dfrac{3}{\sqrt{6}}$

25. $\dfrac{3}{\sqrt{10}}, \dfrac{1}{\sqrt{10}}, 3, \dfrac{1}{3}, \sqrt{10}, \dfrac{\sqrt{10}}{3}$

26. $\frac{9}{41}, \frac{40}{41}, \frac{9}{40}; \frac{9}{41}, -\frac{40}{41}, -\frac{9}{40}$

28. $\dfrac{5}{\sqrt{61}}, \dfrac{6}{\sqrt{61}}, \dfrac{5}{6}; -\dfrac{5}{\sqrt{61}}, -\dfrac{6}{\sqrt{61}}, \dfrac{5}{6}$

29. Sine and cosine are never greater than 1 or less than -1.

Exercise 3-1   *Page 37*

1. 505.5 ft, 506.5 ft   **2.** 28.25 lb, 28.35 lb
4. 0.365 in., 0.375 in.   **5.** 7.1595 cm, 7.1605 cm
7. (*a*) 6440, (*b*) 6400, (*c*) 6000
8. (*a*) 1620, (*b*) 1600, (*c*) 2000
10. (*a*) 0.206, (*b*) 0.21, (*c*) 0.2
11. (*a*) 0.999, (*b*) 1.0, (*c*) 1
13. (*a*) 274,000, (*b*) 270,000, (*c*) 300,000
14. (*a*) 140, (*b*) 140, (*c*) 100
16. (*a*) 2.50×10^3, (*b*) 2.5×10^3, (*c*) 3×10^3
17. 5.8   **19.** 1.95   **20.** 8.8   **22.** 154
23. 24.6   **25.** 270   **26.** 12   **28.** 12

Exercise 3-2   *Page 39*

1. 0.4173   **2.** 0.8541   **4.** 0.4950   **5.** 1.134
7. 0.5543   **8.** 1.402   **10.** 0.9980   **11.** 0.9945
13. 8° 00′   **14.** 3° 30′   **16.** 47° 30′   **17.** 38° 20′
19. 30° 50′   **20.** 27° 50′   **22.** 55° 20′   **23.** 64° 50′
25. 54° 20′   **26.** 50° 40′

Exercise 3-3   *Page 41*

1. 0.3159   **2.** 0.5502   **4.** 0.2189   **5.** 0.8626
7. 0.9468   **8.** 0.8518   **10.** 2.302   **11.** 0.5681
13. 1.034   **14.** 1.479   **16.** 1.322   **17.** 0.9999

19. 0.5509	**20.** 3.250	**22.** 37° 18′	**23.** 59° 32′
25. 64° 02′	**26.** 13° 55′	**28.** 27° 56′	**29.** 85° 16′
31. 68° 23′	**32.** 10° 36′	**34.** 18° 10′	**35.** 47° 34′
37. 41° 48′	**38.** 87° 35′	**40.** 5° 13′	**41.** 5° 24′
43. 64° 42′			

Exercise 3-4   *Page 45*

1. 20°	**2.** 62°	**4.** 80°	**5.** 38°
7. 46°	**8.** 66°	**10.** $\dfrac{\pi}{6}$	**11.** $\dfrac{2\pi}{5}$
13. 0.4540	**14.** −0.9703	**16.** −0.2493	**17.** −1.192
19. −0.5200	**20.** −0.5299	**22.** 5.671	**23.** −0.5000

Exercise 3-5   *Page 48*

The values given are for sin, cos, and tan in that order.

1. (*a*) A: $\frac{4}{5}, \frac{3}{5}, \frac{4}{3}$ B: $\frac{3}{5}, \frac{4}{5}, \frac{3}{4}$

   (*b*) A: $\frac{15}{17}, \frac{8}{17}, \frac{15}{8}$ B: $\frac{8}{17}, \frac{15}{17}, \frac{8}{15}$

   (*c*) A: $\dfrac{5}{\sqrt{74}}, \dfrac{7}{\sqrt{74}}, \dfrac{5}{7}$ B: $\dfrac{7}{\sqrt{74}}, \dfrac{5}{\sqrt{74}}, \dfrac{7}{5}$

2. A: $\frac{5}{13}, \frac{12}{13}, \frac{5}{12}$ B: $\frac{12}{13}, \frac{5}{13}, \frac{12}{5}$

4. A: $\frac{35}{37}, \frac{12}{37}, \frac{35}{12}$ B: $\frac{12}{37}, \frac{35}{37}, \frac{12}{35}$

5. A: $\dfrac{3}{\sqrt{58}}, \dfrac{7}{\sqrt{58}}, \dfrac{3}{7}$ B: $\dfrac{7}{\sqrt{58}}, \dfrac{3}{\sqrt{58}}, \dfrac{7}{3}$

7. A: $\dfrac{7}{\sqrt{113}}, \dfrac{8}{\sqrt{113}}, \dfrac{7}{8}$ B: $\dfrac{8}{\sqrt{113}}, \dfrac{7}{\sqrt{113}}, \dfrac{8}{7}$

8. cos 55°	**10.** cot 22°	**11.** tan 39°	**13.** csc $\dfrac{\pi}{5}$
14. cos $\dfrac{\pi}{6}$	**16.** 20°	**17.** 90°	**19.** 23°
20. $\dfrac{7\pi}{48}$	**22.** $\dfrac{5\pi}{144}$	**23.** $\dfrac{\pi}{42}$	

Exercise 3-6   *Page 51*

13. $2 + \sqrt{3}$ **14.** $-\dfrac{3 + \sqrt{3}}{4}$ **16.** $\dfrac{\sqrt{6} + \sqrt{2}}{4}$ **17.** $\dfrac{\sqrt{6} - \sqrt{2}}{4}$

19. $\dfrac{1}{\sqrt{2}}, -\dfrac{1}{\sqrt{2}}, -1$ **20.** $-\dfrac{1}{\sqrt{2}}, -\dfrac{1}{\sqrt{2}}, 1$ **22.** $-\dfrac{\sqrt{3}}{2}, -\dfrac{1}{2}, \sqrt{3}$

23. $-\dfrac{1}{\sqrt{2}}, \dfrac{1}{\sqrt{2}}, -1$ **25.** $\dfrac{1}{2}, -\dfrac{\sqrt{3}}{2}, -\dfrac{1}{\sqrt{3}}$

Exercise 3-7   Page 54

1. $B = 55°, b = 31, c = 38$ **2.** $A = 50°, a = 39, c = 51$
4. $B = 64°, a = 13, b = 27$ **5.** $B = 37°, a = 56, c = 70$
7. $A = 54°, B = 36°, c = 100$ **8.** $A = 47°, B = 43°, c = 6.8$
10. $A = 53°, B = 37°, c = 0.78$ **11.** $A = 28°, B = 62°, c = 96$
13. $B = 54° 40', b = 354, c = 434$
14. $A = 34° 20', a = 143, c = 254$
16. $B = 46° 30', b = 24.3, c = 33.6$
17. $A = 39° 50', B = 50° 10', c = 5.46$
19. $A = 54° 10', B = 35° 50', b = 31.1$
20. $A = 50° 50', B = 39° 10', a = 2.01$
22. $A = 39° 48', a = 1920, b = 2305$
23. $A = 38° 15', B = 51° 45', c = 6622$
25. $A = 39° 30', b = 0.6991, c = 0.9059$
26. $B = 35° 42', a = 7415, c = 9137$
28. $A = 39° 03', B = 50° 57', c = 6.856$
29. Angles $52° 00'$, sides 64.1   **31.** 18.0 ft
32. 141 in.   **34.** 129 in., 139 in.

Exercise 3-8   Page 58

1. (a) 84 ft, (b) 24 ft   **2.** 390 ft   **4.** 170 ft   **5.** 430 ft
7. 125 ft   **8.** 110 ft   **10.** 22.3 ft   **11.** 602 ft
13. 554 ft   **14.** 530 ft   **16.** 150 ft, 62 ft   **17.** 5.75 ft

19. 50 ft   **20.** $\dfrac{a}{\cot \theta - \cot \phi}$ ft   **22.** 750 ft

23. (a) $\dfrac{a}{\cot \phi - \cot \theta}$, where θ is the angle of depression of the more distant

                 point.

    (b) $\dfrac{a}{\cot \phi + \cot \theta}$

25. 726 ft   **26.** $25°$

Chapter 3   Review exercise   Page 60

1. 2.159   **2.** 0.3242   **4.** $41° 34'$
5. $79° 54'$ **7.** $63°$ **8.** $55°$

10. $42°$ **11.** -0.7771 **13.** 0.2126

14. $\frac{5}{13}, \frac{12}{13}, \frac{5}{12}$ **16.** $\dfrac{2\sqrt{10}}{7}, \dfrac{3}{7}, \dfrac{2\sqrt{10}}{3}$

17. $-\dfrac{\sqrt{3}}{2}, -\dfrac{1}{2}, \sqrt{3}$ **19.** 110 miles

Exercise 4-1   *Page 68*

1. (*a*) $\cos \theta = \sin \theta \cot \theta$, (*b*) $\sin \theta = \dfrac{\cos \theta}{\cot \theta}$

2. (*a*) $\sin \theta = \cos \theta \tan \theta$, (*b*) $\cos \theta = \dfrac{\sin \theta}{\tan \theta}$

4. (*a*) $\tan \theta = \pm \sqrt{\sec^2 \theta - 1}$, (*b*) $\sec \theta = \pm \sqrt{1 + \tan^2 \theta}$
5. (*a*) $\csc \theta = \pm \sqrt{1 + \cot^2 \theta}$, (*b*) $\cot \theta = \pm \sqrt{\csc^2 \theta - 1}$
7. II, III   **8.** I, III

In problems 10 through 19 the answers are given in the order sin, cos, tan, cot, sec, csc.

10. $\frac{4}{5}, \frac{3}{5}, \frac{4}{3}, \frac{3}{4}, \frac{5}{3}, \frac{5}{4}$ **11.** $-\frac{12}{13}, \frac{5}{13}, -\frac{12}{5}, -\frac{5}{12}, \frac{13}{5}, -\frac{13}{12}$

13. $-\dfrac{2}{3}, -\dfrac{\sqrt{5}}{3}, \dfrac{2}{\sqrt{5}}, \dfrac{\sqrt{5}}{2}, -\dfrac{3}{\sqrt{5}}, -\dfrac{3}{2}$ **14.** $\frac{5}{13}, \frac{12}{13}, \frac{5}{12}, \frac{12}{5}, \frac{13}{12}, \frac{13}{5}$

16. $-\dfrac{\sqrt{3}}{\sqrt{5}}, \dfrac{\sqrt{2}}{\sqrt{5}}, -\dfrac{\sqrt{3}}{\sqrt{2}}, -\dfrac{\sqrt{2}}{\sqrt{3}}, \dfrac{\sqrt{5}}{\sqrt{2}}, -\dfrac{\sqrt{5}}{\sqrt{3}}$

17. $-\dfrac{\sqrt{7}}{3}, -\dfrac{\sqrt{2}}{3}, \dfrac{\sqrt{7}}{\sqrt{2}}, \dfrac{\sqrt{2}}{\sqrt{7}}, -\dfrac{3}{\sqrt{2}}, -\dfrac{3}{\sqrt{7}}$

19. $\dfrac{2}{\sqrt{13}}, \dfrac{3}{\sqrt{13}}, \dfrac{2}{3}, \dfrac{3}{2}, \dfrac{\sqrt{13}}{3}, \dfrac{\sqrt{13}}{2}$

20. $\cos \theta = \pm \sqrt{1 - \sin^2 \theta}$, $\tan \theta = \pm \dfrac{\sin \theta}{\sqrt{1 - \sin^2 \theta}}$, $\csc \theta = \dfrac{1}{\sin \theta}$

22. $\sin \theta = \dfrac{\pm \sqrt{\sec^2 \theta - 1}}{\sec \theta}$, $\cos \theta = \dfrac{1}{\sec \theta}$, $\tan \theta = \pm \sqrt{\sec^2 \theta - 1}$

Exercise 4-4   *Page 79*

1. $\dfrac{\sqrt{6} - \sqrt{2}}{4}$ **2.** $\dfrac{\sqrt{6} + \sqrt{2}}{4}$ **4.** $\dfrac{\sqrt{2} - \sqrt{6}}{4}$

5. $-\dfrac{\sqrt{6} + \sqrt{2}}{4}$ **7.** $\dfrac{\sqrt{6} - \sqrt{2}}{4}$ **8.** $\dfrac{\sqrt{2} - \sqrt{6}}{4}$

13. $\cos\theta$ **14.** $-\sin\theta$ **16.** $\cos\theta$

17. $-\sin\theta$ **19.** -1 **20.** $\frac{1}{2}$

22. $\dfrac{\sqrt{3}}{2}$ **23.** $\cos 5A$ **25.** $\sqrt{3}\sin A$

26. $\sqrt{2}\cos A$

28. $\cos(A-B) = \frac{4}{5}$, $\cos(A+B) = \frac{44}{125}$

29. $\cos(A-B) = \frac{13}{85}$, $\cos(A+B) = -\frac{77}{85}$

Exercise 4-5   Page 83

1. $\sin 75° = \dfrac{\sqrt{6}+\sqrt{2}}{4}$, $\tan 75° = 2+\sqrt{3}$

2. $\sin(-15°) = \dfrac{\sqrt{2}-\sqrt{6}}{4}$, $\tan(-15°) = \sqrt{3}-2$

4. $\sin 105° = \dfrac{\sqrt{6}+\sqrt{2}}{4}$, $\tan(105°) = -2-\sqrt{3}$

5. $\sin 225° = -\dfrac{\sqrt{2}}{2}$, $\tan 225° = 1$

7. $\sin 165° = \dfrac{\sqrt{6}-\sqrt{2}}{4}$, $\tan 165° = -2+\sqrt{3}$

8. $\sin(-285°) = \dfrac{\sqrt{6}+\sqrt{2}}{4}$, $\tan(-285°) = 2+\sqrt{3}$

14. 1 **16.** $\dfrac{\sqrt{2}}{2}$ **17.** $\sqrt{3}$ **19.** $\sin 5\theta$ **20.** $\sqrt{3}\sin\theta$

22. $\tan 11A$ **23.** $\cos\theta$ **25.** $\sin\theta$ **26.** $-\cos\theta$

28. $\sin\theta$ **35.** $\frac{56}{25}$, $\frac{56}{33}$, Q I; $-\frac{16}{65}$, $-\frac{16}{63}$, Q IV

37. $\frac{36}{85}$, $-\frac{36}{77}$, Q II; $\frac{84}{85}$, $\frac{84}{13}$, Q I

38. $\dfrac{-2\sqrt{5}}{25}$, $-\dfrac{2}{11}$, Q IV; $\dfrac{2\sqrt{5}}{5}$, -2, Q II

Exercise 4-6   Page 88

1. 0, -1 **2.** $\dfrac{\sqrt{3}}{2}$, $\dfrac{1}{2}$, $\sqrt{3}$ **4.** $\sin 88°$

5. $\cos 96°$ **7.** $\cos 80°$ **8.** $\frac{1}{2}\sin 160°$

10. $\frac{1}{2}\cos 6x$ **11.** $\frac{120}{169}$, $-\frac{119}{169}$, $-\frac{120}{119}$, Q II

13. $\frac{240}{286}$, $-\frac{161}{289}$, $-\frac{240}{161}$, Q II

14. $\dfrac{-4\sqrt{2}}{9}, \dfrac{7}{9}, \dfrac{-4\sqrt{2}}{7}$, Q IV

16. $\frac{1}{2}\sqrt{2-\sqrt{3}}, -\frac{1}{2}\sqrt{2+\sqrt{3}}, 2-\sqrt{3}$

17. $\frac{1}{2}\sqrt{2-\sqrt{2}}, \frac{1}{2}\sqrt{2+\sqrt{2}}, \sqrt{2}-1$

19. $\frac{1}{2}\sqrt{2-\sqrt{2}}, -\frac{1}{2}\sqrt{2+\sqrt{2}}, 1-\sqrt{2}$

20. $\frac{1}{2}\sqrt{2+\sqrt{2}}, -\frac{1}{2}\sqrt{2-\sqrt{2}}, -1-\sqrt{2}$

22. $\sin 20°$ **23.** $\sin 70°$ **25.** $\tan 27°$ **26.** $\tan 32.5°$

28. $2\sin^2 110°$

29. $\dfrac{3}{\sqrt{10}}, \dfrac{1}{\sqrt{10}}, 3$ **31.** $\dfrac{\sqrt{30}}{6}, -\dfrac{\sqrt{6}}{6}, -\sqrt{5}$

32. $\frac{1}{6}\sqrt{3}, \dfrac{\sqrt{33}}{6}, \frac{1}{11}\sqrt{11}$ **34.** $\dfrac{\sqrt{14}}{4}, \dfrac{\sqrt{2}}{4}, \sqrt{7}$

Exercise 4-7   Page 91

1. $\sin 62° + \sin 18°$ **2.** $\sin 65° + \sin 35°$

4. $\frac{1}{2}(-\cos 62° + \cos 10°)$ **5.** $\sin 3\theta + \sin \theta$

7. $\frac{1}{2}(\sin 86° - \sin 8°)$ **8.** $\frac{1}{2}(\sin 58° - \sin 24°)$

10. $\frac{1}{2}(-\cos 5\theta + \cos 3\theta)$ **11.** $2\sin 20° \cos 2°$

13. $2\cos 12.5° \cos 7.5°$ **14.** $-2\sin 47\frac{1}{2}° \sin 2\frac{1}{2}°$

16. $2\cos 5x \cos x$ **17.** $-2\sin 5x \sin x$

19. $2\sin 37\frac{1}{2}° \sin 2\frac{1}{2}°$ **20.** $-2\cos^2 36° \sin^2 5°$

22. $\dfrac{1+\sqrt{3}}{2}$ **23.** $\dfrac{\sqrt{3}+\sqrt{2}}{4}$

25. $\dfrac{\sqrt{2}}{2}$ **26.** $\dfrac{\sqrt{6}}{2}$

Chapter 4   Review exercise   Page 95

1. $\sin\theta = -\dfrac{\sqrt{15}}{4}, \tan\theta = \sqrt{15}, \cot\theta = \dfrac{1}{\sqrt{15}}, \sec\theta = -4,$

$\csc\theta = -\dfrac{4}{\sqrt{15}}$

2. $\sin\theta = \frac{4}{5}, \tan\theta = -\frac{4}{3}, \cot\theta = -\frac{3}{4}, \sec\theta = -\frac{5}{3}, \csc\theta = \frac{5}{4}$

8. $\cos(A-B) = -\frac{33}{65}, \cos(A+B) = \frac{63}{65}$

10. $\tan(A+B) = \frac{21}{220}, \tan(A-B) = \frac{171}{140}$

11. $\sin (A + B) = \dfrac{13}{5\sqrt{10}}$, $\cos (A + B) = \dfrac{9}{5\sqrt{10}}$, $\tan (A + B) = \dfrac{13}{9}$, Q I

$\sin (A - B) = -\dfrac{1}{\sqrt{10}}$, $\cos (A - B) = -\dfrac{3}{\sqrt{10}}$, $\tan (A - B) = \dfrac{1}{3}$,

Q III

13. $\sin 2A = -\frac{24}{25}$, $\cos 2A = \frac{7}{25}$, $\tan 2A = -\frac{24}{7}$, Q IV

14. $\sin \dfrac{1}{2}\theta = \dfrac{2}{\sqrt{5}}$, $\cos \dfrac{1}{2}\theta = \dfrac{1}{\sqrt{5}}$, $\tan \dfrac{1}{2}\theta = 2$, Q I

16. $\sin 66° + \sin 16°$ **17.** $\sin 60° + \sin 20°$

19. $\cos 9\theta + \cos \theta$ **20.** $\frac{1}{2}(\cos 9\theta + \cos 3\theta)$

22. $2 \sin 23° \cos 5°$ **23.** $-2 \sin 4x \sin 2x$

25. $2 \cos 40° \cos 15°$

Exercise 5-1   *Page 110*

1. $\dfrac{\pi}{2}$, 1, -1 **2.** 4π, 3, -3 **4.** $\dfrac{\pi}{3}$, 4, -4

5. $\dfrac{2\pi}{5}$, $\dfrac{1}{2}$, $-\dfrac{1}{2}$ **7.** $\dfrac{\pi}{2}$ **8.** 3π

10. $\dfrac{6\pi}{5}$ **11.** $\dfrac{\pi}{4}$

Chapter 5   Review exercise   *Page 110*

1. 2π, 2, -2 **2.** 4π, 3, -3 **4.** 2π **5.** $\dfrac{2\pi}{3}$, 1, -1

Exercise 6-1   *Page 116*

1. $\left\{\dfrac{7\pi}{6}, \dfrac{11\pi}{6}\right\}$ **2.** $\{\pi\}$ **4.** $\left\{\dfrac{5\pi}{6}, \dfrac{11\pi}{6}\right\}$

5. $\left\{\dfrac{4\pi}{3}, \dfrac{5\pi}{3}\right\}$ **7.** $\left\{\dfrac{\pi}{4}, \dfrac{3\pi}{4}\right\}$ **8.** $\left\{\dfrac{\pi}{6}, \dfrac{11\pi}{6}\right\}$

10. $\left\{\dfrac{\pi}{2}\right\}$ **11.** $\left\{\dfrac{\pi}{8}, \dfrac{3\pi}{8}, \dfrac{9\pi}{8}, \dfrac{11\pi}{8}\right\}$ **13.** $\{38° \ 56', \ 321° \ 4'\}$

14. $\left\{\dfrac{\pi}{8}\right\}$ **16.** $\left\{0, \dfrac{\pi}{3}, \dfrac{2\pi}{3}, \pi, \dfrac{4\pi}{3}, \dfrac{5\pi}{3}\right\}$

17. $\{56° \ 19', \ 236° \ 19'\}$

19. $\left\{\dfrac{\pi}{3}, \dfrac{2\pi}{3}, \dfrac{4\pi}{3}, \dfrac{5\pi}{3}\right\}$ **20.** $\left\{\dfrac{\pi}{3}, \dfrac{2\pi}{3}, \dfrac{4\pi}{3}, \dfrac{5\pi}{3}\right\}$

22. $\left\{\dfrac{\pi}{2}, \pi, \dfrac{3\pi}{2}\right\}$ **23.** $\{0, \pi\}$ **25.** $\left\{\dfrac{\pi}{3}, \dfrac{2\pi}{3}\right\}$

26. $\left\{\dfrac{\pi}{3}, \dfrac{4\pi}{3}\right\}$ **28.** $\left\{\dfrac{\pi}{3}, \pi, \dfrac{5\pi}{3}\right\}$ **29.** $\left\{\dfrac{2\pi}{3}, \dfrac{4\pi}{3}\right\}$

31. $\{14°\ 29', 165°\ 31', 270°\}$ **32.** No solution

34. $\{35°\ 47', 125°\ 47', 215°\ 47', 305°\ 47'\}$

35. $\{33°\ 41', 135°, 213°\ 41', 315°\}$

37. $\{71°\ 34', 135°, 251°\ 34', 315°\}$ **38.** $\{41°\ 49', 138°\ 11'\}$

40. No solution   **41.** $\{206°\ 00', 334°\ 00'\}$

Exercise 6-2   *Page 119*

1. $\left\{\dfrac{\pi}{6}, \dfrac{5\pi}{6}, \dfrac{3\pi}{2}\right\}$ **2.** No solution   **4.** $\left\{\dfrac{\pi}{3}, \dfrac{5\pi}{3}\right\}$

5. $\left\{\dfrac{\pi}{3}, \dfrac{5\pi}{3}\right\}$ **7.** No solution   **8.** $\{63°\ 26', 135°, 243°\ 26', 315°\}$

10. $\left\{\dfrac{2\pi}{3}, \dfrac{4\pi}{3}\right\}$ **11.** $\left\{\dfrac{3\pi}{2}\right\}$ **13.** $\{90°, 216°\ 52'\}$

14. $\left\{0, \dfrac{2\pi}{3}\right\}$ **16.** $\left\{0, \dfrac{\pi}{2}\right\}$ **17.** $\{0°, 250°\ 32'\}$

19. $\{57°\ 40', 159°\ 12'\}$ **20.** $\{36°\ 52', 270°\}$

22. $\left\{0, \dfrac{\pi}{4}, \dfrac{3\pi}{4}, \pi, \dfrac{5\pi}{4}, \dfrac{7\pi}{4}\right\}$ **23.** $\left\{0, \dfrac{\pi}{2}, \pi, \dfrac{3\pi}{2}\right\}$

25. $\{14°\ 2', 90°, 194°\ 2', 270°\}$ **26.** $\{0, \pi\}$

28. $\{26°\ 34', 206°\ 34'\}$ **29.** $\left\{\dfrac{\pi}{2}\right\}$

31. $\left\{\dfrac{\pi}{6}, \dfrac{\pi}{2}, \dfrac{5\pi}{6}, \dfrac{3\pi}{2}\right\}$ **32.** $\left\{\dfrac{\pi}{2}, \dfrac{7\pi}{6}, \dfrac{3\pi}{2}, \dfrac{11\pi}{6}\right\}$

34. $\left\{0, \dfrac{\pi}{3}, \dfrac{2\pi}{3}, \pi, \dfrac{4\pi}{3}, \dfrac{5\pi}{3}\right\}$ **35.** $\left\{\dfrac{\pi}{6}, \dfrac{\pi}{2}, \dfrac{5\pi}{6}, \dfrac{7\pi}{6}, \dfrac{3\pi}{2}, \dfrac{11\pi}{6}\right\}$

37. $\left\{\dfrac{\pi}{6}, \dfrac{5\pi}{6}, \dfrac{7\pi}{6}, \dfrac{11\pi}{6}\right\}$ **38.** No solution

40. $\{47°\ 3', 132°\ 57', 227°\ 3', 312°\ 57'\}$ **41.** $\{90°, 216°\ 52'\}$

43. $\left\{\dfrac{\pi}{12}, \dfrac{5\pi}{12}\right\}$ **44.** $\left\{\dfrac{\pi}{12}, \dfrac{17\pi}{12}\right\}$

Exercise 6-3 *Page 127*

1. 0   **2.** $\dfrac{\pi}{2}$ **4.** 0   **5.** 0

7. $-\dfrac{\pi}{2}$ **8.** π **10.** $\dfrac{\pi}{6}$ **11.** 0.34

13. $\dfrac{5\pi}{6}$ **14.** $-\dfrac{\pi}{4}$ **16.** $\dfrac{2\pi}{3}$ **17.** $\dfrac{\pi}{4}$

19. $\dfrac{\pi}{2} - 0.41$ **20.** $\dfrac{\pi}{2} - 1.25$ **22.** $0.50 - \dfrac{\pi}{2}$ **23.** -0.20

25. $\dfrac{\sqrt{3}}{2}$ **26.** $-\sqrt{3}$ **28.** $\dfrac{1}{\sqrt{2}}$ **29.** $\sqrt{2}$

31. -3 **32.** $-\frac{3}{4}$ **34.** $\frac{1}{3}$ **35.** $\dfrac{3\sqrt{7}}{8}$

37. $\dfrac{-4\sqrt{2}}{7}$ **38.** $\dfrac{\sqrt{5}}{20}$ **40.** $\sqrt{\dfrac{4 + \sqrt{7}}{8}}$ **41.** $\dfrac{3 - \sqrt{5}}{2}$

43. 0.7104   **44.** 0.9249   **46.** -1.052

Exercise 6-4 *Page 130*

11. $\frac{56}{65}$ **13.** $-\frac{63}{16}$ **14.** $\dfrac{4\sqrt{3} - 3}{10}$ **16.** $2x\sqrt{1 - x^2}$

17. $y\sqrt{1 - x^2} + x\sqrt{1 - y^2}$ **19.** $-\sqrt{1 - x^2}$ **20.** $\dfrac{1}{\sqrt{x^2 + 1}}$

22. 1   **23.** $x = \dfrac{1}{\sqrt{5}}$ **25.** $x = \frac{1}{6}$

26. $\left\{ -\dfrac{1}{\sqrt{5}}, \dfrac{1}{\sqrt{5}} \right\}$ **28.** $\dfrac{2}{\sqrt{5}}$ **29.** $x = \pm\dfrac{1}{\sqrt{2}}$

31. $\{-1 \le x \le 1\}$ **32.** $\{-1 \le x \le 1\}$

34. $\{x \mid x \text{ is a real number}\}$ **35.** $\{-1 < x < 1\}$

37. $\{x \mid x \text{ is a real number}\}$ **38.** $\{-1 \le x \le 1\}$

40. $\{-1 \le x \le 1\}$ **41.** $\{x \mid x \text{ is a real number}\}$

43. $\{x \mid x \text{ is a real number}\}$ **44.** $\{-1 \le x < 1\}$

46. $\{x \mid x \le -1 \text{ and } x \ge 1\}$ **47.** $\{-1 \le x \le 1\}$

Chapter 6 Review exercise *Page 131*

1. $\left\{ \dfrac{\pi}{3}, \dfrac{4\pi}{3} \right\}$ **2.** $\left\{ \dfrac{\pi}{3}, \dfrac{2\pi}{3} \right\}$ **4.** $\{218° \, 56'\}$

5. $\left\{\dfrac{3\pi}{2}\right\}$ **7.** $\left\{\dfrac{\pi}{6}, \dfrac{5\pi}{6}\right\}$

8. $\{29°\ 19',\ 105°\ 41',\ 209°\ 19',\ 285°\ 41'\}$

10. $\left\{\dfrac{\pi}{6}, \dfrac{\pi}{2}, \dfrac{5\pi}{6}, \dfrac{3\pi}{2}\right\}$ **11.** $\left\{\dfrac{7\pi}{6}, \dfrac{11\pi}{6}\right\}$

13. 0.42 **14.** 1.26 **16.** $\dfrac{\sqrt{6}}{3}$

17. $\dfrac{\sqrt{8}}{3}$ **19.** $\frac{1}{3}$ **20.** $-\sqrt{3}$

22. $\frac{33}{65}$ **23.** 1

Exercise 7-1   *Page 135*

1. $\log_2 64 = 6$ **2.** $\log_5 125 = 3$ **4.** $\log_7 7 = 1$
5. $\log_6 1 = 0$ **7.** $\log_{81} 3 = \frac{1}{4}$ **8.** $\log_4 \frac{1}{8} = -\frac{3}{2}$
10. $\log_{3/2} \frac{4}{9} = -2$ **11.** $\log_{4/3} \frac{27}{64} = -3$ **13.** $3^0 = 1$
14. $7^1 = 7$ **16.** $8^0 = 1$ **17.** $2^3 = 8$
19. $(\frac{1}{3})^{-2} = 9$ **20.** $3^{-3} = \frac{1}{27}$ **22.** $(\frac{1}{6})^{-2} = 36$
23. $5^{-2} = \frac{1}{25}$ **25.** $x = 100$ **26.** $x = 4$
28. $x = \frac{1}{4}$ **29.** $x = 1$ **31.** $x = \frac{81}{16}$
32. $x = \frac{27}{8}$ **34.** $a = \frac{1}{64}$ **35.** $a > 0$
37. $a = 512$ **38.** $a = 4$ **40.** $a = \frac{27}{64}$
41. $y = 2$ **43.** $y = 3$ **44.** $y = -\frac{1}{3}$
46. $y = \frac{3}{2}$ **47.** $y = -3$

Exercise 7-2   *Page 138*

1. $\log 152 + \log 64 + \log 81$ **2.** $2 \log 43 + \log 7.2$
4. $\frac{1}{3}\log 38 + \log 49 + \log 8.51$ **5.** $\log 4.9 + \log 7.5 - \log 56 - \log 94$
7. $\frac{1}{2}(\log 83 + \log 48 - \log 51)$ **8.** $\frac{1}{2}\log 521 - \log 73 - \log 55$
10. $\frac{1}{4}(\log 68 + \log 126 - \log 79 - \log 66)$

11. $\log 4\pi r^3$ **13.** $\log \pi r^2 h$ **14.** $\log \dfrac{x^5}{2x - 3}$

16. $\log \dfrac{\sqrt[3]{62}}{5(4)^{3/2}}$ **17.** $\log \dfrac{70(726)^2}{91\sqrt{793}}$ **19.** 1.0791

20. 1.2552 **22.** 0.8406 **23.** 0.5187
25. -0.8518 **26.** -0.2273

Exercise 7-3   *Page 143*

1. 1	**2.** 1	**4.** 0	**5.** 2	**7.** 1
8. 3	**10.** -1	**11.** -2	**13.** -4	
14. -5	**16.** -1	**17.** -3	**19.** 1	

20. 8   **22.** 1.4997   **23.** 0.8603
25. 4.6739   **26.** 1.9053   **28.** 7.9138 $-$ 10
29. 8.7634 $-$ 10   **31.** 2.8820   **32.** 2.9031
34. 9.8633 $-$ 10   **35.** 5.8176   **37.** 0.4970
38. 1.6669   **40.** 2.8775   **41.** 9.9510 $-$ 10
43. 4.6841   **44.** 5.3349   **46.** 1.4556
47. 2.9321   **49.** 1.3014   **50.** 9.3477 $-$ 10
52. 2.9464   **53.** 6.3444 $-$ 10   **55.** 5.8808 $-$ 10
56. 5.6594   **58.** 5.6374   **59.** 5.3659 $-$ 10
61. 1.9087 $-$ 10   **62.** 3.9087 $-$ 10

Exercise 7-4   *Page 145*

1. 1640   **2.** 1.87   **4.** 333
5. 6.06×10^8 **7.** 27,100   **8.** 106,000
10. 4.215   **11.** 22.22   **13.** 409.3
14. 6.333   **16.** 0.009856   **17.** 0.0007445
19. 0.13   **20.** 0.002165   **22.** 210.3
23. 5196   **25.** 402.5   **26.** 2205
28. 1393   **29.** 3,536,000

Exercise 7-5   *Page 147*

1. 2.325   **2.** 51.41   **4.** 3.073
5. 0.1079   **7.** 7398   **8.** 134.0
10. 9,078,000   **11.** 20,770   **13.** 0.9662
14. 93.72   **16.** 3.479   **17.** 10.70
19. 2.835   **20.** 1.588   **22.** 0.5014
23. 0.3159   **25.** 0.03734   **26.** 0.4207
28. 3.728   **29.** 0.02253

Exercise 7-6   *Page 149*

1. $x = 3$ **2.** $x = 1$ **4.** $x = -2$
5. $x = 3$ **7.** $x = 2$ **8.** $x = -2$
10. $x = 1.771$ **11.** $x = 1.302$ **13.** $x = 1.931$

14. $x = 0$ **16.** $x = 9.692$ **17.** $x = -18.14$

19. $x = -7.128$ **20.** $x = 0.2870$ **22.** $x = 4$

23. $x = 2$ **25.** $x = 1 + \sqrt{3}$ **26.** $x = 3$

28. $x = \sqrt{29}$ **29.** $x = 1 + 2\sqrt{2}$ **31.** $x = 2 + \sqrt{101}$

Exercise 7-7   *Page 151*

1. 4.094   **2.** 4.970   **4.** 5.701   **5.** 4.620

7. -1.738 **8.** -0.3521 **10.** -1.610 **11.** 1.102

Exercise 7-8   *Page 152*

1. $9.6121 - 10$ **2.** $9.8688 - 10$ **4.** 0.4755

5. $9.6887 - 10$ **7.** $9.9966 - 10$ **8.** $9.9744 - 10$

10. $9.1764 - 10$ **11.** $9.9885 - 10$ **13.** 0.5115

14. 0.4899   **16.** $9.6386 - 10$ **17.** $58° \, 46'$

19. $43° \, 9'$ **20.** $59° \, 54'$ **22.** $17° \, 10'$

23. $45° \, 4'$ **25.** $25° \, 39'$ **26.** $56° \, 6'$

28. $9° \, 23'$ **29.** $24° \, 38'$ **31.** $17° \, 56'$

32. $84° \, 3'$ **34.** $15° \, 52'$ **35.** $14° \, 59'$

Chapter 7   Review exercise   *Page 153*

1. $\log_{10} 100 = 2$ **2.** $\log_{3/4} \frac{27}{64} = 3$

4. $3^5 = 243$ **5.** $2^0 = 1$

7. $5^2 = 25$ **8.** $\frac{1}{2}(\log_{10} 74 + \log_{10} 83)$

10. $\log_{10} 4 + \frac{1}{2}\log_{10} 2 - \log_{10} 3 - \frac{1}{2}\log_{10} 5$

11. $9.5401 - 10$ **13.** 3.8178

14. 6734   **16.** $x = 4$

17. $x = 5$ **19.** $x = 7$

20. $x = \dfrac{-1 + \sqrt{57}}{4}$

Exercise 8-1   *Page 158*

1. $C = 71°$, $a = 51$, $b = 67$ **2.** $B = 87°$, $a = 14$, $c = 27$

4. $A = 98°$, $b = 6.2$, $c = 4.6$ **5.** $A = 60°$, $b = 11$, $c = 25$

7. $A = 59° \, 50'$, $a = 109$, $c = 117$ **8.** $A = 47° \, 10'$, $a = 221$, $b = 230$

10. $A = 69° \, 50'$, $a = 3.62$, $c = 2.57$

11. $A = 37° 57'$, $b = 491.0$, $c = 559.0$
13. $A = 49° 34'$, $a = 60.80$, $b = 78.74$
14. $A = 58° 38'$, $b = 687.8$, $c = 739.3$
16. 228 ft   **17.** 20 ft
19. 4500 ft, 6000 ft, balloon is 3500 ft high.
20. 27 ft

Exercise 8-2   *Page 163*

1. $B = 31°$, $C = 54°$, $c = 41$
2. $B = 33°$, $C = 54°$, $c = 32$
4. No solution
5. $A = 62°$, $C = 70°$, $c = 33$; $A' = 118°$, $C' = 14°$, $c' = 8.7$
7. $A = 100°$, $B = 48°$, $a = 5.7$; $A' = 16°$, $B' = 132°$, $a' = 1.4$
8. $A = 55°$, $B = 54°$, $a = 8.5$
10. $A = 90°$, $C = 53°$, $a = 60$; $A' = 16°$, $C' = 127°$, $a' = 17$
11. $A = 37° 10'$, $B = 42° 40'$, $a = 124$
13. $A = 31° 48'$, $C = 99° 40'$, $c = 4554$
14. $A = 125° 53'$, $B = 21° 33'$, $a = 1.329$ **16.** No solution
17. $A = 41° 19'$, $B = 47° 10'$, $a = 4319$ **19.** No solution
20. $A = 48° 1'$, $B = 63° 28'$, $a = 247.9$ **22.** $50°$
23. Two possibilities: 26 miles or 9.2 miles
25. $54°$, 6.3 ft

Exercise 8-3   *Page 167*

1. $A = 51°$, $B = 76°$, $c = 4.1$ **2.** $B = 43°$, $C = 75°$, $a = 9.1$
4. $A = 23°$, $B = 38°$, $c = 3.4$ **5.** $B = 25°$, $C = 128°$, $a = 14$
7. $B = 20°$, $C = 31°$, $a = 0.64$ **8.** $B = 75°$, $C = 33°$, $a = 0.69$
10. $A = 54°$, $B = 49°$, $C = 77°$ **11.** $A = 42°$, $B = 52°$, $C = 86°$
13. $A = 29° 40'$, $B = 44° 20'$, $C = 106° 00'$ **14.** No solution
16. 27 ft and 48 ft   **17.** 24.8 ft

Exercise 8-4   *Page 170*

1. 51   **2.** 51   **4.** 5.8   **5.** 3200
7. 1330   **8.** 21.9   **10.** 106
11. Two possibilities: 2500 and 269
13. 480   **14.** 30.5

Chapter 8   Review exercise   *Page 171*

1. $a = 4.2$ **2.** $c = 97$ **4.** $c = 648$
5. $C = 35° \, 3'$ or $144° \, 57'$ **7.** $c = 33$
8. $C = 76°$ **10.** 30 miles   **11.** 87° and 93°
13. $B = 50°$, $C = 95°$, $c = 21$; $B' = 138$, $C' = 15°$, $c' = 5.4$
14. 19 ft and 35 ft

Exercise 9-1   *Page 182*

1. 260 lb, 39°, 51°   **2.** S32° E, 390 lb
4. 170 lb, 270 lb   **5.** 1700 ft/sec, 1200 ft/sec
7. 1600 lb, 1700 lb   **8.** 33°
10. 1300 lb   **11.** 270 mph, 79°
13. 100 mph, 90 mph   **14.** 170 mph, 50 mph
16. 250 lb, 97 lb   **17.** 120 lb, 260 lb
19. 700 lb, 490 lb

Exercise 9-2   *Page 186*

1. 230 lb, 310 lb   **2.** 1.6 mph, 2.9 mph
4. 260 mph, 300 mph   **5.** N 46° E, 3.0 mph
7. Current 1.6 mph, speed 2.9 mph   **8.** 100 lb
10. 143°, 90°, 126°   **11.** 470 miles, 172°
13. 129°, 260 miles   **14.** N 44° E, 6.5 ft/sec

Exercise 9-3   *Page 191*

1. $-2i + 8j$, $6i - 2j$ **2.** $5i + 4j$, $5i - 4j$
4. $i + 2j$, $11i + 12j$ **5.** $\frac{3}{5}i + \frac{4}{5}j$

7. $\dfrac{1}{\sqrt{17}}i - \dfrac{4}{\sqrt{17}}j$ **8.** $\dfrac{1}{\sqrt{5}}i + \dfrac{2}{\sqrt{5}}j$

10. $\frac{8}{17}i + \frac{15}{17}j$ **11.** $(4, -1)$
13. $(\frac{3}{2}, -1)$ **14.** $(\frac{3}{2}, \frac{1}{2})$
16. $(0, -1)$, $(3, 1)$ **17.** $(-3, 4)$, $(0, 5)$
19. $(-\frac{4}{3}, -1)$, $(\frac{1}{3}, 3)$

Exercise 9-4   *Page 196*

1. $A \cdot B = 0$, $\cos \theta = 0$ **2.** 56, $\frac{56}{65}$
4. 16, $\frac{8}{17}$ **5.** $a = \frac{2}{3}$

7. $37°$ (nearest degree), $\cos PRQ = -\dfrac{\sqrt{10}}{10}$, $\cos PQR = 0.26$

8. $45°$, $90°$, $45°$ 10. $\frac{42}{13}$

11. $\sqrt{2}$

Chapter 9   Review exercise   *Page 197*

2. 170 lb, 210 lb   4. $81°$, $137°$, $142°$

5. $(2, -\frac{1}{2})$

Exercise 10-1   *Page 203*

1. $-6 - 3i$ 2. $-2i$ 4. $-16 - 8i$
5. $1 + 9i$ 7. $x = -4, y = 3$ 8. $x = 4, y = -\frac{3}{2}$
10. $x = \frac{5}{2}, y = -\frac{5}{2}$ 11. $x = -4, y = -7$ 13. $10 + 9i$
14. $-1 + 4i$ 16. $-1 + 16i$ 17. 7
19. $-4 + 6i$ 20. $11 - 15i$ 22. $8 - 6i$
23. $(5\sqrt{3} - \sqrt{2})i$ 25. $19 + 7i$ 26. $14 + 23i$
28. 41   29. $2 - 26i$ 31. $3 - i$
32. $28 + 4i$ 34. $-2 - 4i$ 35. $\frac{9}{25} + \frac{12}{25}i$
37. $\frac{11}{37} + \frac{8}{37}i$ 38. $\frac{7}{10} - \frac{9}{10}i$ 40. $-\frac{5}{13} - \frac{12}{13}i$
41. $\frac{16}{65} + \frac{63}{65}i$

Exercise 10-2   *Page 208*

1. $3 + 2i$ 2. $6 - 5i$ 4. -5
5. $-4 - 2i$ 7. $-3 + 4i$ 8. $4 + 3i$
10. $\sqrt{2}(\cos 45° + i\sin 45°)$ 11. $2(\cos 225° + i\sin 225°)$
13. $2(\cos 330° + i\sin 330°)$ 14. $4(\cos 30° + i\sin 30°)$
16. $2(\cos 270° + i\sin 270°)$ 17. $3(\cos 180° + i\sin 180°)$
19. $2(\cos 45° + i\sin 45°)$ 20. $10(\cos 300° + i\sin 300°)$
22. $2\sqrt{7}(\cos 300° + i\sin 300°)$ 23. $5(\cos 36° 52' + i\sin 36° 52')$
25. $1 + i\sqrt{3}$ 26. $2\sqrt{3} + 2i$ 28. $-4\sqrt{2} + 4i\sqrt{2}$
29. $-2\sqrt{3} + 2i$ 31. -3 32. $-6i$
34. $-5.35 - 5.94i$

Exercise 10-3   *Page 211*

1. $-3 + 3i\sqrt{3}$ 2. -20 4. $-7\sqrt{2} - 7i\sqrt{2}$

5. $27i$ 7. -2 8. $\dfrac{3\sqrt{2}}{2} + \dfrac{3i\sqrt{2}}{2}$

10. $2\sqrt{2}\,(\cos 345° + i\sin 345°)$
11. $2\sqrt{2}\,(\cos 285° + i\sin 285°)$
13. $20\sqrt{2}\,(\cos 135° + i\sin 135°)$
14. $8(\cos 0° + i\sin 0°)$

16. $\dfrac{\sqrt{2}}{3}(\cos 45° + i\sin 45°)$

17. $2(\cos 270° + i\sin 270°)$

19. $\dfrac{\sqrt{2}}{2}(\cos 105° + i\sin 105°)$

Exercise 10-4   Page 214

1. $9(\cos 68° + i\sin 68°)$ **2.** $8(\cos 75° + i\sin 75°)$
4. $256(\cos 88° + i\sin 88°)$ **5.** $\cos 200° + i\sin 200°$
7. $4(\cos 180° + i\sin 180°)$ **8.** $8(\cos 270° + i\sin 270°)$
10. $64(\cos 180° + i\sin 180°)$ **11.** $1024(\cos 210° + i\sin 210°)$
13. $3(\cos 20° + i\sin 20°)$, $3(\cos 200° + i\sin 200°)$
14. $5(\cos 66° + i\sin 66°)$, $5(\cos 246° + i\sin 246°)$
16. $2(\cos 37° + i\sin 37°)$, $2(\cos 127° + i\sin 127°)$
   $2(\cos 217° + i\sin 217°)$, $2(\cos 307° + i\sin 307°)$
17. $3(\cos 22° + i\sin 22°)$, $3(\cos 112° + i\sin 112°)$,
   $3(\cos 202° + i\sin 202°)$, $3(\cos 292° + i\sin 292°)$
19. $\sqrt[6]{2}\,(\cos 105° + i\sin 105°)$, $\sqrt[6]{2}\,(\cos 225° + i\sin 225°)$,
   $\sqrt[6]{2}\,(\cos 345° + i\sin 345°)$
20. $\cos 30° + i\sin 30°$, $\cos 150° + i\sin 150°$, $\cos 270° + i\sin 270°$
22. $2\sqrt[4]{2}\,(\cos k\cdot 90° + i\sin k\cdot 90°)$, $k = 0, 1, 2, 3$
23. $\cos(18° + k\cdot 72°) + i\sin(18° + k\cdot 72°)$, $k = 0, 1, 2, 3, 4$
25. $-32\sqrt{3} - 32i$
26. -32

28. $\dfrac{1}{2} + \dfrac{\sqrt{3}}{2}i$

29. $-\dfrac{1}{2} + \dfrac{\sqrt{3}}{2}i$

31. $i,\ -\dfrac{\sqrt{3}}{2},\ -\dfrac{1}{2}i,\ \dfrac{\sqrt{3}}{2} + \dfrac{1}{2}i$

32. $-2i,\ \sqrt{3} + i,\ -\sqrt{3} + i$
34. $2^{3/4}(1 + i),\ 2^{3/4}(-1 + i),\ 2^{3/4}(-1 - i),\ 2^{3/4}(1 - i)$
35. $0.9511 + 0.3090i,\ i,$
   $-0.9511 + 0.3090i,\ -0.5878 - 0.8090i,$
   $0.5878 - 0.8090i.$

Exercise 10-5 *Page 218*

1. $(3,-300°)$, $(-3,240°)$, $(-3,-120°)$
2. $(4,-240°)$, $(-4,-60°)$, $(-4,300°)$
4. $(2,-120°)$, $(-2,60°)$, $(-2,-300°)$
5. $(2,-270°)$, $(-2,270°)$, $(-2,-90°)$
7. $(4,180°)$, $(-4,0°)$, $(-4,360°)$
8. $(5,270°)$, $(-5,90°)$, $(-5,-270°)$
10. $(2,0°)$, $(-2,180°)$, $(2,360°)$
11. $(3,315°)$, $(3,-45°)$, $(-3,135°)$

Exercise 10-7 *Page 224*

1. $(0,3)$ 2. $(1,1)$ 4. $(0,0)$ 5. $(0,2)$
7. $(3,90°)$ 8. $(5,0°)$ 10. $(0,\theta)$, $\theta \geq 0°$ 11. $(2,315°)$
13. $(4,300°)$ 14. $(2,240°)$ 16. $r = 5 \sec \theta$

17. $r = -4 \csc \theta$ 19. $r = \dfrac{6}{2 \cos \theta - 3 \sin \theta}$

20. $r = 2$ 22. $x^2 + y^2 = 25$
23. $x + y = 0$ 25. $y = -6$ 26. $x = 2$

Chapter 10 Review exercise *Page 224*

1. $x = \frac{3}{4}, y = \frac{3}{2}$ 2. $x = -28, y = 12$
4. $1 - i$ 5. $14 - 8i$
7. $-3 - 6i$ 8. $\frac{3}{13} + \frac{11}{13}i$
10. $(6,0)$ 11. $(6,8)$
13. $(\frac{1}{2},\frac{1}{2})$ 14. $(0,\frac{1}{2})$
16. $4(\cos 60° + i \sin 60°)$ 17. $2(\cos 315° + i \sin 315°)$
19. $1 + i$ 20. $2 + 3i \sqrt{3}$
22. $1, -\dfrac{1}{2} + i\dfrac{\sqrt{3}}{2}, -\dfrac{1}{2} - i\dfrac{\sqrt{3}}{2}$

23. $(4,-300°)$, $(-4,-120°)$, $(-4,240°)$
25. $(2,360°)$, $(-2,180°)$, $(-2,-180°)$

Exercise A-1 *Page 229*

1. $A = 34°, B = 87°, c = 27$ 2. $A = 118°, B = 35°, c = 38$
4. $B = 10°, C = 9°, a = 79$ 5. $A = 92°, C = 42°, b = 45$
7. $B = 55° 10', C = 33° 10', a = 110$

   8. $A = 30° \ 20'$, $B = 60° \ 30'$, $c = 822$
10. $B = 17° \ 00'$, $C = 19° \ 40'$, $a = 116$
11. $B = 64° \ 13'$, $C = 30° \ 54'$, $a = 7.821$
13. $A = 39° \ 29'$, $B = 21° \ 57'$, $c = 0.4707$
14. $B = 47° \ 51'$, $C = 99° \ 34'$, $a = 1338$

Exercise A-2   *Page 232*

   1. $A = 20°$, $B = 32°$, $C = 128°$ **2.** $A = 44°$, $B = 61°$, $C = 75°$
   4. $A = 30°$, $B = 37°$, $C = 113°$ **5.** $A = 51°$, $B = 39°$, $C = 90°$
   7. $A = 43° \ 20'$, $B = 50° \ 40'$, $C = 86° \ 00'$
   8. $A = 48° \ 40'$, $B = 53° \ 30'$, $C = 77° \ 50'$
10. $A = 115° \ 00'$, $B = 32° \ 00'$, $C = 33° \ 10'$
11. $A = 117° \ 49'$, $B = 37° \ 27'$, $C = 22° \ 44'$

index